PLANT CLOSINGS

AND WORKER DISPLACEMENT

The Regional Issues

Marie Howland

1988

W. E. UPJOHN INSTITUTE for Employment Research

Library of Congress Cataloging-in-Publication Data

Howland, Marie, 1950–
   Plant closings and worker displacement : the regional issues /
Marie Howland.
       p.   cm.
   Bibliography: p.
   Includes index.
   ISBN 0-88099-063-5.   ISBN 0-88099-062-7 (pbk.)
   1. Plant shutdowns—United States.   2. Employees, Dismissal of—
United States.   I. Title.
   HD5708.55.U6H69   1988
338.6'042—dc19   88-17032
                                                               CIP

THE INSTITUTE, a nonprofit research organization, was established on July
1, 1945. It is an activity of the W. E. Upjohn Unemployment Trustee Corporation, which was formed in 1932 to administer a fund set aside by the late
Dr. W. E. Upjohn for the purpose of carrying on "research into the causes
and effects of unemployment and measures for the alleviation of unemployment."

The facts presented in this study and the observations and viewpoints express-
ed are the sole responsibility of the author. They do not necessarily represent
positions of the W. E. Upjohn Institute for Employment Research.

ii

iii

AUTHOR

Dr. Marie Howland is an associate professor in the Institute for Urban Studies at the University of Maryland, College Park. She is also a regular consultant for the Urban Institute in Washington, D.C. In 1981, she received her Ph.D. from the Department of Urban Studies and Planning at the Massachusetts Institute of Technology. In addition to her work on plant closings, she has published articles on the response of city and regional economies to national business cycles, the role of property taxes in influencing the location of new firms, and the impact of capital subsidies on rural development.

ACKNOWLEDGEMENTS

I am grateful to both the Institute for Urban Studies at the University of Maryland and the Urban Institute in Washington, D.C. for institutional support while I was writing portions of this manuscript. In particular, George Peterson's support was crucial in the initiation of this project. George Peterson, along with Margaret Dewar, Mel Levin, Patricia Sawyer, Phil Shapira, Paul Flaim, Yvonne Benningfield and Kristin Sutherlin read portions, or in some cases all, of the chapters and made valuable suggestions. Discussions with Marc Bendick and Ken Corey were helpful in formulating several arguments, and chapter 5 reflects many revisions suggested by Mead Over, whose assistance with the econometrics is greatly appreciated. Steve Sawyer was helpful in locating and making sense out of the energy data used in chapter 4, and Todd Girdler was an excellent research assistant who meticulously assembled most of the data. Both Bennett Harrison and William Wheaton are responsible for stimulating my interest in the topic of plant closings while I was a graduate student at the Massachusetts Institute of Technology.

My husband, Michael Wattleworth, was a continuous source of encouragement, and as always Matthew and Colin Wattleworth were my most enthusiastic supporters.

A special debt is owed to Lou Jacobson of the Upjohn Institute who exhibited continued confidence and interest in the topic, and who contributed many valuable ideas and suggestions and forced me to strengthen many arguments along the way. Finally, I want to thank those who provided financial assistance, the Upjohn Institute, the Computer Science Center at the University of Maryland, the Division of Behavioral and Social Sciences at the University of Maryland, and the U.S. Department of Housing and Urban Development who funded the purchase of the Dun and Bradstreet Data. To all of the above I owe thanks, and, of course, I alone am responsible for any remaining errors.

Contents

Tables

Figures

1
Plant Closings
The Problem and Current Policy

In the 25 years immediately following World War II, the United States economy experienced moderate industrial and regional restructuring. The decline of the manufacturing sector and the movement of capital and population from parts of the North to most of the South and West took place amidst vibrant national economic growth and relative cyclical stability, and, consequently, went unnoticed in most parts of the country. In the last 18 years, these trends have accelerated. Not only is restructuring now occurring more rapidly than in previous decades, but it is taking place in an environment of weaker national growth and greater cyclical instability. In this context, it is not surprising that the loss of manufacturing jobs and the decline of many of the older industrialized regions have caught the attention of policymakers and academics, and that issues surrounding worker displacement have risen near the top of the domestic policy agenda. More and more often, manufacturing workers seem to be the victims of plant closings and permanent layoffs in regions and industries where employment opportunities are limited.

The aggregate numbers show that in recent years, job loss in some locations and industries has been significant. Between 1981 and 1986, the manufacturing sector lost 1.2 million jobs. This decline was not shared equally by all regions and industries. During the same period, for example, the Mid-Atlantic region lost 0.6 million jobs, while the East South Central gained 16,000 and the Mountain states gained 31,000 (U.S. Department of Commerce 1985 and 1988). There were also structural changes within manufacturing, with the skill-extensive, high-wage industries experiencing the largest employment losses. The steel industry lost almost 230,000 jobs; at the same time, office and computing

1

machine manufacturing gained 94,000. While manufacturing employment declined in the U.S., the service sector grew by 4.5 million jobs (U.S. Department of Commerce 1982-83 and 1988).

Large overall employment losses in manufacturing hide the fact that a far larger number of jobs were lost and workers displaced. For example, with the net loss of 1.2 million manufacturing jobs, estimates from the U.S. Bureau of Labor Statistics are that approximately 2.6 million manufacturing workers, with previously stable work histories, were involuntarily and permanently laid off between 1981 and 1986 (U.S. Department of Labor 1986).

These permanent job losses can be traumatic for the individuals affected. Numerous studies have documented the psychological and financial costs of displacement and have found that the consequences can be devastating for individuals and communities. Many victims of plant closures and mass layoffs experience prolonged periods of unemployment, and, if they find new jobs, they often accept a dramatic cut in their standard of living. Medical studies have shown that dislocated workers exhibit rates of anomie, depression, alcoholism, heart disease and suicide that far exceed the rates in the general population. The problems are particularly acute for many older dislocated workers because they have had a stable work history (and consequently have rusty job search skills) and a high-paying job and comfortable life style which cannot be duplicated. Furthermore, they are less geographically mobile and often are perceived by employers as being less retrainable than younger workers (Gordus, Jarley, and Ferman 1981). Communities can also face difficult adjustments after a plant closing or mass layoff. When a large employer closes or downsizes, local governments can be hit with a declining tax base, a rising tax burden, and reductions in the quality of services.

The purpose of this study is to look behind net employment changes to (1) examine the relationship between regional employment shifts and plant closures, and (2) draw the implications of that relationship for displaced worker policy. More specifically, this study explores four questions.

• Is regional economic restructuring responsible for high rates of plant closures and permanent layoffs in the less competitive regions?

- What are the characteristics of the enterprises most likely to close? Does a plant's age, size, or status as a headquarters, independent, branch, or subsidiary affect the probability of a closure?
- To what extent do adverse local economic conditions, such as high wages, utility costs, and taxes; a unionized labor force; or shrinking local market influence the probability a plant will close?
- Do workers displaced in an expanding local labor market experience large postdisplacement financial losses, or is economic dislocation a problem specific to the Rust Belt economies?

The intention of this study is both to increase our conceptual understanding of the process of economic decline and job loss, and to contribute to the ongoing debate on displaced worker policy.

Plant Closures and the Dynamics of Regional Change

In regional economics and regional planning literature, plant closures are widely believed to be both cause and consequence of industrial restructuring. When an industry shifts location, high rates of plant closures in the less competitive region are assumed to coincide with high start-ups in the low-cost or higher-revenue regions (see for example, Bluestone and Harrison 1982; Markusen 1985; Mansfield 1985). Underlying causes of employment shifts to the South and West, or the so-called Sun Belt, are changes in regional production costs, technological change within industries, growing union strength among regional workforces, intensified industrial competition, and declining local markets.

Changes in relative regional production costs can alter the least-cost location calculus for an industry. For example, the slowdown in European immigration into New England in the 1920s caused the price of unskilled labor to rise in the northern states. Southern and southwestern locations, with abundant low-skilled labor from rural areas, became more attractive to assembly line manufacturing. Second, technological change in an industry's production process can free many industries to seek lower input costs in new locations. For example, with the development of steam power in the 1830s, the textile industry was no longer tied to the inland rivers of New England, and many firms relocated toward the coast and the South to cut labor costs. Third, restructuring may also be capitalists' response to growing militancy or organization on the part of labor (Bluestone and Harrison 1982; Storper and Walker 1984). Firms may close branches, relocate operations and open new plants in the non-union areas as a strategy for keeping in check labor's demand for higher wages and other benefits. Fourth, intensifying international competition is cutting into the profit margins of many domestic companies, forcing them to take new cost-cutting measures. One such measure may be a move to lower-cost production sites. Markusen (1985) argues persuasively that the steel industry would have decentralized earlier, to be near markets and to reduce transportation costs, if not for the industry's oligopolist structure, market power, and high profits. Prior to foreign competition, the steel industry did not have an incentive to tighten costs. Finally, the loss in local markets has led to the regional shift of market-oriented manufacturing out of the Frost Belt.

Manufacturing industries in the process of decentralizing are generally beyond their earliest stages of product development, and are consequently no longer anchored to a region-specific, specialized labor force. If not tied to a natural resource, they are free to relocate, and if they operate in a competitive environment, they have the incentive to seek out their most profitable location.

In order to understand the process of regional employment shifts and its relationship to plant closures, this study examines the components of employment change in three decentralizing industries, metalworking machinery, electronic components, and motor vehicles. These industries are intended to be representative of many manufacturing industries shifting out of the northern states to the South and West.

The components of employment change include plant start-ups, closings, relocations and net on-site expansions. At the outset, there were three possible findings.

> (1) Industry plant closure rates and rates of worker dislocation are highest in the regions where the industry is exhibiting the greatest decline.
> (2) Dislocation occurs at even rates in all regions of the country, but worker displacement is not a problem in growing regions because new jobs absorb the laid-off workers.

(3) Displacement occurs evenly in all regions, but workers
do not move easily into new firms. Consequently, displaced
workers are forced to accept dramatic losses in income, even
in growing regions.

In the first and second cases, worker displacement is an issue confined
primarily to declining economies. In the third instance, worker displacement is a national concern.

The findings support case 3 above, with two reservations. First, in
industries experiencing dramatic declines in employment and where plant
start-up rates are already low, as with motor vehicles, permanent layoffs
appear to be highest in the declining regions. In other words, case 1
applies. The second exception is that the youngest, best-educated, white
and female workers displaced in labor markets exhibiting strong employment growth appear to make the transition to a new job with relatively
small or no income losses. In other words, for these workers, case 2
applies.

Current Public Policy to Assist Displaced Workers

Policymakers have acknowledged the high costs of job loss for the
stable worker, but a coherent federal policy is still in the formative stages.
At present, there are four policies which address some aspects of the
displaced worker problem: import protections for selected domestic industries, Trade Adjustment Assistance, the Job Training Partnership
Act, and plant closing legislation.

The U.S. restricts imports of a number of commodities, with the explicit purpose of protecting domestic jobs in those industries. At present, textiles, apparel, some leather products, steel, large motorcycles,
and small trucks are protected by import duties, voluntary restraints,
or import quotas. Several additional industries are protected by antidumping duties, including semiconductors and steel. An antidumping
case can be brought when foreign producers are found to either sell
their product in U.S. markets at a lower price than they charge at home
or sell their product in U.S. markets below cost. Trade protection is
the only major policy in force that attempts to slow structural shifts and

inhibit job loss, and even this policy has become less common with time (Goldstein 1986). Increasingly, uncompetitive industries are forced to improve the quality of their product, cut costs, or experience major revenue losses that lead to employment declines. The emphasis of displaced worker policy has switched away from protecting jobs through trade restraints to providing some adjustment assistance for workers after layoff.

One such policy and a second form of assistance to displaced workers is Trade Adjustment Assistance (TAA). TAA provides supplemental financial assistance and a small amount of job search and relocation assistance to workers who lose their job as a result of the liberalization of trade. The legislation states that TAA is to be made available to workers when imports "contributed importantly" to their separation. The justification for this assistance, which supplements unemployment insurance, is that it will reduce political resistance to trade liberalization and provide partial income maintenance to the workers who face the greatest barriers to reemployment. Because of questions about the equity of TAA, and findings that many recipients are later reemployed by the same company, there is growing congressional pressure to phase out TAA and shift additional resources to the Job Training Partnership Act (JTPA Title III), the third program to assist displaced workers.

JTPA Title III makes aid available to all dislocated workers, not just those affected by trade. These funds are flexible, but the most common uses are to set up plant-based job search and training assistance. The Reagan administration has voiced support for this program, and Congress increased funding from $223 million in 1985 to $287 million in fiscal year 1988. JTPA's performance evaluations noted delays in getting the money to a closing plant, shortages in funds for hard hit areas, and lack of knowledge about the program on the part of displaced workers (U.S. Congress 1986).

A fourth set of policies to assist displaced workers falls in the category of plant closing legislation. This legislation has focused on assisting all workers affected by plant closures. Presumably, the legislation excludes workers who lose their jobs through cutbacks, because a layoff is clearly permanent when a plant closes, while it is not easy to determine whether layoffs from a continuing plant are permanent or cyclical.

The proposed legislation has been designed to both minimize the personal losses and community costs of plant closures and to reallocate the social costs of displacement to those responsible for the layoff decision.

In some states, protections for workers have already been adopted. Maine and Wisconsin have enacted legislation requiring employers to give workers 60 days notice before a merger, liquidation, relocation, or plant closure. Connecticut, Maine, Massachusetts, Maryland and Delaware have passed laws to promote employee ownership, and California, Michigan, New Jersey, New York, and Pennsylvania have passed laws providing loans and/or loan guarantees and technical assistance for worker buyouts of closing plants (Rosen, Klein, and Young 1986, pp. 253-254). Other states have introduced, but not enacted, bills to require prenotification or to assist workers in buying out their closing plant.

Legislation to shift some of the burden of large scale job losses from workers to firms has also been introduced at the national level. In 1974, Senator Walter Mondale of Minnesota and Congressman William Ford of Michigan introduced the National Employment Priorities Act (NEPA). This original bill died in committee, and was unsuccessfully reintroduced in an altered form every year until 1984. The major components of these bills provided for employee prenotification of a closing, severance pay to separated workers, grants and loans to failing businesses under some circumstances, and economic redevelopment assistance to local governments. In response to strong industry opposition and the bill's failure to get out of committee, Congressman Ford, along with Congressman Conte of Massachusetts, introduced a new bill in 1985 as a stopgap measure. This bill required that workers be given 90 days advance notice of a closing for plants with 50 or more employees, and was designed to aid displaced workers until a blue ribbon commission could evaluate the need for a more comprehensive bill. This last initiative was narrowly defeated on the floor of the House.

In response to a request from Congress, Secretary of Labor William Brock established an economic adjustment and worker dislocation taskforce, made up of representatives of industry, government and academia. The taskforce's January 1987 recommendations call for a merger of the JTPA and TAA programs into a Worker Adjustment Assistance Program, with additional spending for job search, training, and cash assistance. The taskforce could not agree on a law requiring mandatory prenotification of a closing for displaced workers and communities.

In 1988, Congress again introduced displaced worker legislation. This time it passed both Houses as part of the 1988 Omnibus Trade Bill. This bill included $900 million for worker retraining, continuation of TAA, and 60 days advance notice of a closing for firms with 100 employees or more. In mid-1988, President Reagan vetoed the bill. While the House of Representatives mustered enough votes to override the veto, an override failed in the Senate.

In contrast to the U.S. experience, the largest industrialized countries outside of the United States have already instituted some benefits for displaced workers. In Sweden, Great Britain, and West Germany, corporations are legally obligated to give advance notice before a closing and to negotiate the closing with their employee unions or workers' councils. The extent of other protections, such as severance payments, varies by country (U.S. Congress 1983, pp. 113-163).

In the United States, most structural adjustments are allowed to take place with little public sector intervention, leaving workers to bear a substantial share of the costs. In part, our reluctance to adopt a coherent and successful policy to deal with displaced workers can be attributed to a conservative political trend, but the effort to implement displacement legislation has also been hampered by a lack of understanding about the relationship between plant closures and economic growth, about the causes of plant closings, and ultimately about the most efficient and effective way to help displaced workers. This study attempts to address these issues.

Organization and Major Findings

The study is organized into five additional chapters. Chapter 2 describes the Dun and Bradstreet (D&B) data, used to carry out the analysis in chapters 3 and 4. This data set includes establishment level information for the years 1973, 1975, 1979, and 1982 for three industries, metalworking machinery, electronic components, and motor vehicles. From these data, we can estimate regional and central city-suburban employment changes, subdivided by plant closures, start-ups, relocations, and on-site contractions and expansions. Chapter 2 explores the reliability and shortcomings of these data.

Chapter 3 examines the hypothesis that plant closures within industries are more frequent events in the economies where the industry exhibits the greatest decline. The chapter begins by describing the regional distribution of employment and employment trends for metalworking machinery, electronic components, and motor vehicles. All three industries are decentralizing from the industrialized North to the Sun Belt states. Metalworking machinery and motor vehicles are also decentralizing within regions, whereas electronic components employment is growing fastest in central cities. Regional employment shifts are analyzed by their components of growth, and finally the chapter tests whether, holding establishment characteristics constant, a plant is more likely to close in declining than in growing economies.

Four points characterize the process of regional restructuring for the three industries. (1) Employment shifts to the Sun Belt are not explained by relatively high rates of plant closures in the Frost Belt, but by high rates of job creation in the Sun Belt and, to a limited extent, by plant migrations from the Frost Belt to the Sun Belt. This same pattern also holds for intraregionals shifts in employment. Plant closure rates are relatively even across central city, suburb, and nonmetropolitan areas, and uneven rates of growth are explained by spatial variations in job creation through plant start-ups and expansions. (2) After holding constant a plant's status as a branch, subsidiary, headquarters or independent, and its size, there is no evidence that plant closure rates vary by location. (3) In the metalworking machinery and electronic components industries, the number of displaced workers is substantially higher in the older, industrialized states only because these industries are concentrated in those states, not because of higher closure rates. (4) There is some evidence for the 1975-79 period, in the motor vehicle industry only, that job loss through plant closures was greater in regions with greater industry decline. These results suggest that industrial decline may initially take the form of falling rates of job creation, which fail to compensate for losses occurring through an average rate of plant closures (as was found for the metalworking and electronic components industries). When an industry's rate of job creation is already very low, industrial decline is furthered by a rising rate of job loss through plant shutdowns.

Chapter 4 examines the hypothesis that high wages, utility costs, and taxes, relatively large increases in wages and utility costs, a unionized labor force, degree of import penetration, and/or shrinking market demand are responsible for plant closures. While some evidence from the metalworking machinery industry shows that the relocation of establishments is more likely out of unionized states than nonunion states, we find no evidence that the variables commonly believed to affect plant closures do, in fact, have an impact. Instead, plant closure decisions appear to reflect the strategies and idiosyncracies of individual firms.

The limited cross-industry results are similar. There is no evidence that plant closure rates are higher in slow-growth than in expanding industries. The only variable to consistently influence the probability a plant will close is the plant's status as a subsidiary or branch. Branch plants and subsidiaries are between 8 and 32 percent more likely to close than headquarters or single plant operations.

Even with equal rates of plant closures, and, for metalworking machinery and electronic components, permanent job loss in all regions of the country, worker displacement may still be a regional problem, specific to the economies where new job creation is not sufficiently strong to absorb workers laid off by plant closures and permanent layoffs. The analysis in chapter 5 relies on a January 1984 Bureau of Labor Statistics survey of displaced workers to identify the labor market conditions under which displacement is a problem. Specifically, we test the hypothesis that worker displacement is still primarily a problem concentrated in declining economies, because workers laid off in labor markets with high rates of job creation successfully make the transition to a new job.

Chapter 5 argues that workers displaced in economies where the industry of displacement is growing are unemployed shorter periods of time and with smaller financial losses than workers displaced in areas where their industry of displacement is declining; that many displaced workers do not move easily into growing industries after displacement; and that many workers, especially those who are older and less educated, experience large reductions in living standards after displacement, even when they are displaced in a growing local labor market. We therefore conclude that worker displacement is a national issue, especially for older, less educated workers, and not a concern specific to the Rust Belt economies.

These findings indicate that the commonly accepted view of the causes of worker displacement is inaccurate and the problems of worker displacement, so evident in the Frost Belt region, are not simply a consequence of regional and industrial restructuring. A worker is as likely to be displaced in a growing region as in a declining region, and for the industries studied, as likely to be displaced from a declining as a growing industry. Structural shifts are, however, responsible for the increasingly high costs of displacement. Many displaced workers do not move easily into new occupations and industries and as a consequence regional realignments mean that the new, compatible jobs are frequently in the wrong location. Furthermore, industrial shifts mean that the skill requirements of newly created jobs do not match those of many displaced workers. The reemployment barriers are, therefore, particularly severe for workers displaced from shrinking industries in stagnating or declining economies.

Chapter 6 explores several policy options for both national policymakers and local economic development officials and argues for increased federal support to assist in the local takeovers of closing branch plants and subsidiaries, and for financial and adjustment assistance, especially for older, less educated displaced workers. The study's findings argue against industrial policy as a means of slowing the pace of worker dislocation, and against concessions in wages, utility bills, and taxes as a strategy for retaining local jobs.

REFERENCES

Bluestone, Barry and Bennett Harrison (1982) *The Deindustrialization of America.* New York: Basic Books.

Goldstein, Judith (1986) "The Political Economy of Trade: Institutions of Protection," *American Political Science Review* 80, 1 (March), pp. 161-184.

Gordus, Jeanne Prial, Paul Jarley, and Louis Ferman (1981) *Plant Closings and Economic Dislocation.* Kalamazoo, Michigan: W.E. Upjohn Institute for Employment Research.

Mansfield, Edwin (1985) *Microeconomics,* shorter fifth edition. New York: W.W. Norton.

Markusen, Ann (1985) *Profit Cycles, Oligopoly, and Regional Development.* Cambridge, Massachusetts: MIT Press.

Rosen, Corey, Katherine J. Klein, and Karen M. Young (1986) *Employee Ownership in America: The Equity Solution.* Lexington, Massachusetts, Lexington Books.

Storper, Michael and Richard Walker (1984) "The Spatial Division of Labor: Labor and the Location of Industries" in Larry Sawers and William Tabb (eds.), *Sunbelt/Snowbelt: Urban Development and Regional Restructuring.* New York: Oxford University Press.

United States Congress (1983) *Testimony before the Subcommittee on Labor Management Relations of the Committee on Education and Labor.* 98th Congress, First Session, HR 2847. Washington, D.C.: Government Printing Office, May 4 and 19.

United States Congress, Office of Technology Assessment (1986) *Plant Closing: Advance Notice and Rapid Response.* Washington, D.C.: Government Printing Office, September.

United States Department of Commerce, Bureau of the Census (1982-83, 1985, and 1988) *Statistical Abstract.* Washington, D.C.: Government Printing Office.

United States Department of Labor, Bureau of Labor Statistics (1979 and 1984) *Employment and Earnings.* Washington, D.C.: Government Printing Office.

United States Department of Labor, Bureau of Labor Statistics (1986), "Reemployment Increases Among Displaced Workers," *News,* October 14.

2

The Dun and Bradstreet Data

The Dun and Bradstreet (D&B) data underlie the analysis in chapters 3 and 4. Chapter 2 examines the reliability of these data by comparing D&B's employment estimates with those of other sources and by evaluating the ways in which the data may bias estimates of plant closings, worker displacement, and employment growth due to start-ups, relocations, and on-site contractions and expansions.

In its role as a credit-rating company, Dun and Bradstreet collects and maintains information on approximately 6 million establishments. This data base, called the Dun's Market Identifiers (DMI) file, includes a Dun's number assigned to each business establishment; the establishment's business address; the number of employees at that location; the business's major Standard Industrial Classification (SIC) at the 4-digit level; and the establishment's status as an independent, headquarters, branch, or subsidiary.

The Urban Institute obtained a subset of the Dun and Bradstreet file that included all establishments listing either SIC 354 (metalworking machinery), SIC 367 (electronic components), or SIC 371 (motor vehicles) as primary, secondary, or tertiary lines of business. To permit an analysis of employment changes over time, the above data were obtained for the peak year of 1973, the recession year of 1975, the peak year of 1979 and the recession year of 1982. The sample includes data on 27,014 distinct establishments in SIC category 354, 14,067 establishments in SIC category 367, and 11,909 establishments in SIC category 371.

These three industries were selected for a number of reasons. First, their growth rates vary. According to the Bureau of Labor Statistics' *Employment and Earnings,* motor vehicles is a stagnant industry, with an annual average rate of employment growth over the 1973-79 period of 0.9 percent. Metalworking machinery is an average-growth industry, exhibiting a 2.2 percent average annual growth rate, and electronic components is a fast-growth industry, with a growth rate of 4.7 percent over the same period. For purposes of comparison, total manufacturing employment grew from 1973 to 1979 at an annual average growth rate of 2.4 percent.

A second reason for selecting these industries is that all three are decentralizing from the industrialized northern states to the Sun Belt. Consequently, we could use the data to examine the process of regional decline.

Third, there is at least anecdotal evidence of a large number of closings among them.

Fourth, the three industries were selected to provide a variety of industry structures. Motor vehicle employment is concentrated in a few large companies, while metalworking machinery employment, in contrast, is dispersed among many relatively small producers.

Fifth, metalworking machinery and electronic components were selected because the establishments in these industries are reasonably well-represented in all of the major census regions of the U.S. (Each region's share of national employment in the three industries is shown in table 3.1 in chapter 3.)

A final reason for this selection of industries is that motor vehicles and machine tools represent basic industries, whereas electronic components is considered part of the high technology sector. These cross-industry differences are more likely to capture variations in plant closure behavior and to permit generalizations from three to many industries.

Creating the Data Set

In order to estimate plant closing rates and employment losses due to plant closings, as well as employment changes due to start-ups, relocations, and expansions or contractions, the four D&B files were merged to create histories for each establishment.

Information from the City Reference File (CRF) of the Bureau of the Census was then added to the merged D&B file. The CRF assigns place descriptions to zip code areas. Thus, each establishment located in a Standard Metropolitan Statistical Area (SMSA) was noted as having a central city or suburban location. The 1977 Census definition of a central city is used here and includes the largest city in an SMSA. One or two additional cities may be identified as a central city on the basis of the following criteria (U.S. Bureau of the Census 1979, p. 25).

1. The additional city or cities must have a population of one-third or more of that of the largest city and a minimum population of 25,000; or

2. The additional city or cities must have at least 250,000 inhabitants. The DMI file records an SMSA code for each establishment, thus allowing us to determine whether a firm is located in or outside an SMSA. Suburbs, for this study, are defined as the area within an SMSA as noted by the DMI file and outside a central city, as noted by the CRF. The 1977 boundaries are used throughout this study, in order to maintain consistency in geographical comparisons.

Coverage

The D&B data are expected to provide reasonably complete coverage for all types of establishments in the manufacturing sector except branch plants. D&B collects these data for credit rating reference, and branches are unlikely to require a credit rating independent from the owning company. Thus D&B has been less assiduous in recording information for branch plants than for headquarters or single-plant firms.

There are two ways that this underreporting of branch plants has been handled by other users. The two major users of the D&B DMI file, to date, are David Birch of the Massachusetts Institute of Technology, and Armington, Odle, and Harris of the Brookings Institution. Both groups worked with the complete data set, including all industries. For headquarters, the data set includes the total number of employees in the company and the total number of employees at that site. For single-plant firms (independents), the two numbers are the same, and for branches the record includes only the number of employees at that site. The Brookings team accounted for the underreporting of branches by assuming the employment totals for the whole company, found in the headquarters record, were accurate. They then created "imputed" records for missing branches. The MIT group, on the other hand, assumed the establishment level data were accurate.

We followed the MIT group's approach because the industry of the headquarters is not a good proxy for the industry of the missing branches. Jacobson (1985) offers further support for our approach. In a comparison of establishment level data between the D&B file for Texas and the Texas unemployment insurance file, he found underreporting of branches by D&B not to be a serious problem. In addition, the Brookings group's method of "imputing" branches overestimates employment in the U.S. because foreign branches of domestic firms are included in the total.

Even though employment totals may underestimate branch plant employment, the D&B data set captures 100 percent of each industry's employment in all regions, with one or two exceptions. Table 2.1 compares total regional employment estimates from the D&B data base with total employment as estimated by the *Census of Manufactures*. The employment estimates from both data sources include only those establishments that regard each respective industry as their first line of business. The ratios shown in table 2.1 are employment totals as estimated by D&B, divided by totals as counted by the Census for 1982. The ratios are disaggregated by region to detect the existence of any regional biases in the undercounting of employment.

For metalworking machinery and motor vehicles, the D&B data estimates are higher than those of the *Census of Manufactures*. One reason to expect lower Census than D&B estimates is that the figures from the former source are aggregated from state data, and the total for several states is suppressed to avoid the disclosure of information about a single firm. For other states, a range of numbers rather than an actual figure is given. When this was the case, we took the midpoint of the range. The D&B data are not subject to the disclosure restrictions of government data sources, and thus may give a more complete count of employment totals. Because of the high coverage, the D&B data are treated in several parts of this study as a population total, rather than as a sample.

The D&B data have several advantages over other data sources. First, establishment level data permit growth rates to be partitioned into the components of growth, including start-ups, closings, expansions and contractions, and migration. As mentioned above, the D&B data on individual establishments are publicly available and confidentiality is not an issue. Second, businesses have a commercial interest in providing accurate information, and consequently are likely to take more care in providing data to D&B than to the U.S. government. In spite of these advantages, the data base has a number of shortcomings, which are addressed in the remainder of this chapter.

Table 2.1
Ratio of Employment Totals from Dun and Bradstreet
to Employment Totals from *Census of Manufactures* 1982, by Region

Region	Metalworking machinery	Electronic components	Motor vehicles
New England	1.42	1.00	1.21
Mid-Atlantic	1.76	.74	1.31
East North Central	1.26	.94	1.19
West North Central	2.91	.73	1.03
South Atlantic	1.17	.76	1.40
East South Central	1.06	1.47	1.46
West South Central	1.33	1.41	1.26
Mountain	1.07	.87	1.39
Pacific	.95	1.07	1.31
Total	1.31	.96	1.23

SOURCES: Numerator=total employment as estimated from the Dun and Bradstreet File, 1982. Denominator=total employment from U.S. Census, *Census of Manufactures, Geographic Area Series, State Books,* 1982.

New England: Maine, Vermont, New Hampshire, Massachusetts, Connecticut and Rhode Island.

Mid-Atlantic: New York, New Jersey, and Pennsylvania.

East North Central: Ohio, Illinois, Indiana, Michigan, and Wisconsin.

West North Central: North Dakota, South Dakota, Minnesota, Iowa, Missouri, Nebraska, and Kansas.

South Atlantic: Delaware, Maryland, District of Columbia, West Virginia, Virginia, North Carolina, South Carolina, Georgia, and Florida.

East South Central: Alabama, Tennessee, Kentucky, and Mississippi.

West South Central: Texas, Louisiana, Arkansas, and Oklahoma.

Mountain: Montana, Wyoming, Idaho, Utah, Nevada, Colorado, Arizona, and New Mexico.

Pacific: California, Oregon, and Washington.

Problems in Estimating Plant Closings
with the D&B Data

As with any data set, especially one not intended for research, there are a number of shortcomings. One shortcoming of the DMI file is that not all establishments are interviewed every year. The data sets acquired from D&B are as they existed on December 31, 1973, December 31, 1975, December 31, 1979, and July 28, 1982, and not all firms were interviewed during the year of the tape's date. Fortunately, the D&B file records the date of establishment interview so that a distribution of interview dates could be calculated. The tables were calculated by region, so as to detect any regional biases in the updating of establishments. The table for metalworking machinery for 1975 is shown as table 2.2. The 1979 and 1982 results and other industry results are similar (see Howland 1983). We report the 1975 results because they are the earliest year for which we have the survey date for each establishment. Since D&B has attempted to improve the accuracy of the data over time, these are reported as baseline estimates. In general, we find no regional bias and, furthermore, that about 80 percent of all establishments are updated in a year. Only about 6 percent of the records are more than two years out of date.

Table 2.2
Percent of Interviews Taken in and Prior to 1975
As Recorded on the 1975 D&B Tape
By Region for Metalworking Machinery
(SIC 354)

	1967-69	1970-72	1973	1974	1975
New England	1	5	6	8	79
Mid-Atlantic	2	5	5	8	80
East North Central	1	4	4	8	83
West North Central	1	5	4	8	82
South Atlantic	1	4	5	10	79
East South Central	0	3	2	10	85
West South Central	1	5	4	8	82
Mountain	1	4	4	9	83
Pacific	1	5	7	10	78

SOURCE: Dun and Bradstreet DMI file, 1975.

The failure to update records may bias the plant closing estimates downwards. Outdated records are probably inactive firms which have done little or no borrowing in the capital market, and we suspect that many of them are out of business. Birch and the MIT group treated outdated records as open and left them in the file. The Brookings group took into account that some of the records may have been closures and dropped out-of-date records from their analysis.

To estimate the extent to which outdated records are closings, we attempted to track down all of the Maryland records that were more than one year out-of-date, using both the telephone directories and the *Maryland Directory of Manufactures.* As stated above, the percentage, and therefore the number, of outdated records is relatively small. The percent of incorrect records is shown in table 2.3.

Table 2.3 shows, for example, there were only five metalworking machinery establishments in Maryland more than two years out-of-date in the 1982 D&B file. The November 1982 and 1984 phone books and the *Maryland Directory of Manufactures* gave no evidence that four of the five companies were still operating, resulting in an error rate of 80 percent. For metalworking machinery records that were one to two years out-of-date, 29 percent of the 14 establishments appeared to be closed. The remaining 71 percent had a current phone number and address. Clearly, the error rate is higher for the more dated records.

Also shown in table 2.3 is the employment error rate. There were 15 employees in the five establishments outdated by more than two years; 73 percent of those jobs were incorrectly recorded as still in existence. The employment error rate is much lower among establishments one to two years out-of-date. These numbers indicate that small establishments are more likely to have outdated records in the D&B file. According to the Maryland phone book and the *Maryland Directory of Manufactures,* all of the out-of-date records that appeared to be closed were independents. The few branch plants in the outdated group were all still in existence.

A second shortcoming of the data base that affects the accuracy of plant closing estimates is that from time to time, D&B purges out-of-date records on establishments they believe to be closed. D&B sends first class letters to each inactive establishment requiring information of their status. When no response is received, D&B assumes the establishment is closed and retires its Dun's number. Some of these eliminated records might actually have been moves rather than closings, and the establishment might never have received D&B's letter.

Table 2.3
Percent of Inaccurately Recorded Establishments
in the D&B File, for Maryland

	Percent of out-of-date establishments incorrectly recorded as open		Percent of recorded closings found to be open		
	(Total number)		(Total number)		
	More than 2 years out-of-date	1-2 years out-of-date	Independents	Branches	All
Metalworking Machinery					
ESTAB	.80	.29	.39	0	.35
	(5)	(14)	(31)	(3)	(34)
EMPLOY	.73	.002			.31
	(15)	(3090)			(662)
Electronic Components					
ESTAB	.88	.17	.30	.19	.27
	(8)	(12)	(50)	(16)	(66)
EMPLOY	.88	.02			.41
	(24)	(1064)			(1819)
Motor Vehicles					
ESTAB	.25	.25	.30	0	.23
	(4)	(12)	(10)	(3)	(13)
EMPLOY	.20	.004			.55
	(5)	(5680)			(150)

SOURCES: Dun and Bradstreet DMI file; Maryland Phone Directories, November 1979, October 1982, October 1983; and the *Maryland Directory of Manufactures*, Department of Economic and Community Development, Annapolis, MD, 1977/78, 1979/80, 1981/82, and 1983/84.

The MIT and Brookings researchers also recognized this problem, but dealt with it in different ways. Birch's group at MIT assumed that the removed firms were still operating and included them in their employment counts. The Brookings group assumed the removed firms were, in fact, closings.

To determine the extent to which purged firms were movers, we again took all the establishments that D&B recorded in Maryland as closing and attempted to track them down through the phone directories and the *Maryland Directory of Manufactures.* A substantial number of establishments were found to be operating after D&B dropped them from the file (table 2.3). Most of the still operating establishments had changed location, with only 2 out of 113 operating at the same address recorded in the D&B file. Unfortunately, the error rate for both establishments and employment is high, ranging from 23 to 35 percent for establishments and 31 to 55 percent for employment.

We disaggregated these error rates by independents and branch plants and found a higher error rate for independents than for branch plants (table 2.3). Because it is possible that a branch could close and other branches of the same company still show up in the Maryland phone book or *Directory of Manufactures,* we made phone calls to branches registered by D&B as closed that had a listing in one of the directories. The error rate in table 2.3 reflects all corrections due to the closing of one branch of a company and the continued operation of sister branches.

The very high error rate is handled in two ways. In the analyses that rely on aggregate totals, such as regional totals, we keep the outdated records in the file to cancel some of the error caused by attributing a closed status to some establishments still in operation. If Maryland is typical of the rest of the nation, the net effect should be a slight overestimate of establishment closings and a larger overestimate of employment lost in plant closings.

In the analysis (in chapters 3 and 4) based on individual observations, the focus is on establishments rather than employment. The error rate for establishments is below that for employment. In addition, we attempt to avoid some of the biases caused by errors in the data by separating branches and independents, since the error rate for branch plants is relatively low. As shown in table 2.3, there were only three branches in all three industries that were incorrectly recorded as closings, and there were no branches whose records were more than two years out-of-date and who were closed. The errors among both the out-of-date records and relocations recorded as closings are concentrated in single-plant operations.

Attributing a closed status to some proportion of movers may give the data a regional bias. Movers are most likely to be growing enterprises that move to obtain more space. Since growing establishments are most likely to be located in growing regions, the plant closing rates in growing regions may be overestimated more than the rates in slow-growth regions. Potential biases are discussed in more detail in the results sections of chapters 3 and 4.

A third factor that may affect closing rates is a change in legal status. D&B's procedure is for acquired establishments to keep the same Dun's number. We did not find any evidence to suggest that recently acquired firms were given new Dun's numbers and thus recorded as a closing and an opening. We therefore assumed that retired numbers were in fact closings.

A fourth problem that might affect estimates of displaced workers, as well as estimates of employment gains or losses due to start-ups, relocations, or expansions and contractions is coding errors in the employment numbers. In the original D&B tape, the number of employees was coded as YXXX, where Y is the number of zeros to be attached to XXX. Any error in coding Y could easily distort employment values by thousands of employees. To check for such errors, a listing of all establishments that experienced employment changes of 1,500 employees or greater between any two years was printed out. There were 1,779 establishments that fell into this category in all three industries. These cases were reviewed by hand to check for coding errors. In some cases the change looked plausible. In one establishment, for example, employment went from 7,500 to 6,000. Such cases were left unchanged.

There were, however, cases where coding errors were obvious. For example, one establishment was recorded as having 300,000 employees in 1975, and 151 employees in 1979. The 1975 value was changed to 300. The number of cases where similar errors were detected and records revised was 5 cases for SIC 354, 17 for SIC 367, and 16 for SIC 371. Where errors were not obvious, the original numbers were left unchanged.

To determine the accuracy of estimates of displaced workers from the D&B data, estimates from this source were compared to the estimates derived from the Current Population Survey (CPS) of Displaced Workers from January 1984. The comparisons are shown in table 2.4. While the time spans of the two data sets do not precisely overlap, the figures are annualized to facilitate comparisons. The D&B estimates cover the period December 1979 to July 1982. The CPS estimates cover the period January 1979 to January 1984.

Table 2.4
Comparison of Estimates of Displaced Workers
Metalworking Machinery and Motor Vehicles
(per annum)

Region	Metalworking machinery		Motor vehicles	
	D&B[a]	BLS[b]	D&B[a]	BLS[b]
New England	2,557	127	282	836
Middle Atlantic	3,436	757	4,378	1,560
East North Central	7,623	4,511	33,123	13,494
West North Central	1,019	—	2,515	2,102
South Atlantic	1,112	337	1,881	427
East South Central	866	—	1,282	1,356
West South Central	627	921	2,881	1,644
Mountain	414	—	78	213
Pacific	1,962	455	3,081	3,313

SOURCES: Dun and Bradstreet DMI file, 1979 and 1982 and U.S. Department of Labor, Bureau of Labor Statistics, Current Population Survey—Supplement on Displaced Workers, January 1984.

a. Includes workers displaced due to plant closings or a relocation out of state, covers the years December 1979 to July 1982.

b. Includes workers displaced due to either plant closings or the failure of a self-operated business, January 1979 to January 1984.

— Sample too small to provide estimates.

Data for electronic components are not included in table 2.4 because the CPS data do not have an equivalent industrial category. It should be noted that the CPS counts are derived from a sample, which is less accurate in instances where the total population is small, such as for metalworking machinery employment in the West North Central, East North Central, and Mountain states. As indicated above, the D&B data tend to overestimate closures. Therefore, the D&B estimates on the number of displaced workers were expected to be higher than those of the BLS, which is often, but not always, the case. Another reason D&B estimates were expected to exceed those of the BLS is that if a worker intended to retire or quit at the same time his or her job was eliminated by a plant closure, a job lost would not represent a displaced worker in the BLS data but would be included in the D&B estimates. Given the differences in the ways the two data sets were assembled, in the dates of their coverage, and between definitions of jobs lost and displacement, the results are reasonably close and suggest that results derived from the D&B data will be meaningful.

Problems in Estimating Start-Ups
and Relocations With the D&B Data

In spite of the problems outlined above, the closing estimates from the D&B file are expected to be more reliable than the start-up estimates. Dun and Bradstreet are constantly expanding their coverage of establishments, especially branch plants, but they do not give a date of start-up for branch plants. Therefore, it is not possible to determine whether a branch added to the file is actually a birth or the addition of a previously existing but unrecorded establishment. A date of inception is reported for headquarters and independents.

The MIT group and the Brookings group approached this problem in different ways. Brookings assumed that all branches added to the file were in fact new start-ups. The only exception to this rule occurred when a newly listed headquarters was added to the file. Then all of its branches were also assumed to be new listings rather than births. The MIT group, on the other hand, used statistics on employment in multiplant enterprises for 1967 to 1977 to estimate the likely level of employment change due to actual branch births in the 1978-80 period. This procedure led to estimates of branch births that were less than half of the estimates of Brookings. Both procedures are problematic. Clearly, Brookings includes as births many branches already in existence. The MIT approach does not take into account the cyclical sensitivity of birth rates.

The estimation of start-ups is further complicated by a serious underreporting of new firm start-ups. It can be several years before a new plant is included in the D&B file. Brookings adjusted for this shortcoming by accepting any newly listed nonbranch business that was established up to five years prior to inclusion in the file as a 1978-80 start-up. To estimate nonbranch births, Birch and the MIT group estimated the rate at which D&B absorbed start-ups over the period 1967 to 1980. The MIT team then calculated "absorption factors" by which recorded births must be multiplied in order to estimate actual births in the "real world." Different absorption factors were calculated for the five major industry groups and for different time intervals. This procedure probably leads to start-up estimates that are too high, due to the implicit assumption that D&B promptness in recording births has not improved between 1969 and 1980 and the employment in recorded births is the same as that in unrecorded births.[1]

Our adjustment to account for underreported start-ups and the inability to distinguish between branch start-ups and the addition of a previously existing branch to the file differs from both of the above approaches. We estimated employment growth rates for the periods 1973-75, 1975-79, and 1979-82 from the D&B data set. To calculate these growth rates we included only those firms that registered a date of birth between 1973-75, 1975-79, or 1979-82, respectively. All other establishments that were added to the file over the three periods were ignored. These employment growth rates were then compared with employment growth in each industry as estimated for the same time period by *Employment and Earnings*. We then assumed that any differences between the two growth rates were due to the underreporting of births by the D&B file.

This procedure is based on two assumptions. First, the DMI file captures the same percentage of start-ups in all locations of the U.S. and during all phases of the business cycle. Second, the divergence between the D&B and *Employment and Earnings* estimates are solely due to unreported start-ups. Using this procedure, we estimate that D&B captures approximately 1 out of 7 new jobs in start-ups in the metalworking machinery industry, 1 out of 11 new jobs in start-ups in the electronic components industry, and 100 percent of new jobs in start-ups in the motor vehicle industry. It seems reasonable to find greater coverage of new start-ups for the slower-growing industries since there are fewer births in those industries to capture.

Another problem arose in attempting to identify each establishment as having either a central city or suburban location. The post office is constantly dividing zip code areas and creating new zip codes. For the zip codes created after 1977, there were no matching location identifiers in the 1977 CRF. Since the majority of these new zip codes were outside SMSAs, the problem was not as serious as it might have been. Out of 32,253 records in 1982, there were 204 establishments located within SMSAs whose zip codes did not have matches in the CRF. These establishments were eliminated from the central city/suburban analysis. The number of cases in this category is sufficiently small (0.6 percent of all establishments in SMSAs) that the elimination of these cases should not distort the results.

A further difficulty arose in attempting to identify intrametropolitan movers. One problem with the DMI data is that many addresses are not legitimate street addresses. Birch (1979, p. 17) estimated that, for all industries, about 20 percent of addresses were illegitimate. They were names of office buildings, industrial parks, shopping plazas, or street intersections. In other cases, addresses were abbreviated in one year and not in another; for example, Skyline *Rd* was reported as the address for one firm in 1975 and Skyline *Road* was reported in 1979. Due to both of these problems, in combination with misspellings, the matching of street addresses to determine movers was impossible. For this reason we decided to match zip codes rather than addresses to identify intrametropolitan movers. Two problems arose. First, firms occasionally use the zip code of the nearest post office rather than the code of the location of their facility. This creates a city bias in identifying the location of establishments. Second, as mentioned above, zip code boundaries changed over time. Consequently, it was difficult to determine whether an establishment was a mover or whether its zip code was redefined. This second problem was resolved through an editing process carried out as follows.

According to the post office, when changes are made in a geographical area's zip code, only the last two numbers of the zip code are affected. In order to distinguish between an actual move and a redefinition of zip codes, a list of firms for which the last two digits changed during interim years and whose addresses were not the same was printed out. Many nonmovers were included in this file. For example, in 1975 Jephco Manufacturing had a zip code of 74135. In 1979 the code was 74112. The address for Jephco Manufacturing was recorded as 3704 E *56th St.* in 1975 and 3704 E *56 St.* in 1979. This establishment is clearly operating in the same location. Establishments such as Jephco were remerged with the file and flagged as nonmovers.

An exception to zip code changes affecting only the final two digits occurred in 1980. During that year, the post office revised the last three or four digits of a number of zip codes. Establishments affected by these changes were accurately recorded as nonmovers.

Clearly, the use of zip codes arbitrarily includes some short moves in the mover file and excludes others, but this should not have a substantial effect on the results. The study will analyze central city to suburban moves, SMSA to nonmetropolitan moves and interregional moves. The number of establishments moving within a zip code district and yet changing type of geographical place should be very small.

A final bias of the data is that it probably underestimates relocations. Many companies, especially larger multiplant firms, often change location by setting up a branch in a new location at the same time they phase out an obsolete plant. Capital equipment at the old site may deteriorate as cash flows generated at that plant are shifted to finance the new plant. When the new plant is well-established, and/or the older plant is no longer profitable, the company finally closes the latter down. While this process may be considered a relocation or "runaway shop," it is counted by D&B as a start-up at the new plant and a closing of the old.

Summary

The Dun and Bradstreet establishment data were not assembled for academic research and therefore are flawed in several respects. In spite of the shortcomings identified above, however, it remains the most complete data source for identifying spatial patterns of plant closings and worker displacement. An understanding of the problems and biases of the data permits a more careful interpretation of the results that follow in chapters 3 and 4.

NOTE

1. See Johnson and Storey (1985) for a good discussion of the differences between the MIT and Brookings adjustments to the D&B data.

REFERENCES

Armington, Catherine and Marjorie Odle (1981) "Associating Establishments into Enterprises for a Microdata File of the U.S. Business Population," *Statistics of Income and Related Record Research.* Internal Revenue Service. Washington, D.C.: Government Printing Office, pp. 71-76.

Birch, David (1979) *The Job Generation Process.* Cambridge, Massachusetts: MIT Program on Neighborhood and Regional Change.

Birch, David and S. McCracken (1983) *The Small Business Share of Job Creation: Lessons Learned from the Longitudinal File.* Cambridge, Massachusetts: MIT Program on Neighborhood and Regional Change.

Harris, Candee (1983) *Small Business and Job Generation: A Changing Economy or Differing Methodologies.* Washington, D.C.: Brookings Institution, February 25.

Howland, Marie (1982) "Using the Dun and Bradstreet Data to Analyze the Effects of Business Fluctuations on Firm Employment," Report submitted to the U.S. Department of Housing and Urban Development, Urban Institute report No. 3165-04. Washington, D.C.: Urban Institute.

Howland, Marie (1983) "Cyclical Effects at the Local Level: A Microeconomic View," Report submitted to U.S. Department of Housing and Urban Development, Urban Institute report No. 3165-07. Washington, D.C.: Urban Institute.

Jacobson, Louis (1985) "Analysis of the Accuracy of SBA's Small Business Data Base," working paper (CNA) 85-1958. Alexandria, Virginia: The Public Research Institute, Center for Naval Analysis, October.

Johnson, Steve and David Storey (1985) "Job Generation—An International Survey," research working paper 1. University of New Castle Upon Tyne, Scotland.

United States Department of Commerce, Bureau of the Census (1977) "Economic Censuses Geocoding: City Reference File Record Layout and Field Definitions," memorandum.

United States Department of Commerce, Bureau of the Census (1979) *Census Geography.* Washington, D.C.: Government Printing Office.

United States Department of Labor, Bureau of Labor Statistics (1973-1976) *Employment and Earnings.* Washington, D.C.: Government Printing Office.

United States Department of Labor, Bureau of Labor Statistics (1977-1981) *Supplements to Employment and Earnings: Revised Establishment Data.* Washington, D.C.: Government Printing Office.

3

Spatial Patterns of Plant Closings, Job Dissolution, and Economic Growth

Chapter 3 uses the Dun and Bradstreet (D&B) data to examine the hypothesis that business closings and worker displacement are more frequent events in slow-growth than fast-growth economies. The factors that discourage the start-up of new firms and branch plants, and lead to economic stagnation or decline, are also expected to accelerate job dissolution. Relatively low regional profit rates caused by high local wages, a unionized labor force, and high utility costs have been found to discourage regional investment (Engle 1974; Carlton 1979; Schmenner 1982; Howland 1985; and Harris 1986). These same factors should also depress relative profit levels in existing establishments, leading to above-average rates of plant closures, runaway shops, and shrinkage in ongoing enterprises.

The hypothesis is tested by (1) comparing regional rates of plant closings with rates of local employment growth, and (2) determining whether plants are more likely to close in the slow-growth regions, when plant age, status, and size are held constant.

Chapter 3 is divided into five sections. The first section describes the regional distribution of employment along with regional employment shifts. The second section presents rates of employment growth for metalworking machinery, electronic components, and motor vehicles by regional and metropolitan/nonmetropolitan economies, and examines the extent to which high rates of plant closure are responsible for economic decline. This section also considers the remaining components of growth (start-ups, relocations, and expansions or contractions) and their role in explaining geographic differences in employment growth. The third section examines the relationship between economic growth and rates of job loss due to plant closings, relocations, and contractions. The fourth section presents regional estimates of the number of

31

displaced workers, and the final section reports the results of logit models testing the impact of regional location on the probability a plant will close, holding constant the plant's size, age, and status as a headquarters, independent, branch, or subsidiary.

As stated in chapter 2, we use data on three 3-digit Standard Industrial Code industries, metalworking machinery (SIC 354), electronic components (SIC 367), and motor vehicles (SIC 371). By disaggregating the data to three specific industries rather than all manufacturing, we avoid some of the complications caused by regional differences in industrial composition. For example, closing rates could be high in a region because all industries close at high rates there, or because of that region's dependence on industries with high closing rates. We are particularly interested in examining how closure rates vary across regions within a single industry. To provide background to the analyses that follow, the first portion of this chapter describes the spatial distribution of employment in the three industries studied and their regional employment shifts.

Regional Employment Distribution and Trends

Table 3.1 shows the distribution and shift of national employment between 1973 and 1979 for metalworking machinery, electronic components, and motor vehicles. The 1973-79 growth rate was selected because both 1973 and 1979 were peak years in the national business cycle. The use of a peak-to-peak growth rate reduces biases in cross-space comparisons caused by variations in subnational business cycles (see for example, Howland 1984; Howland and Peterson 1988). If, for example, we took growth rates using the recession year of 1982 as the end point, the regional shifts out of the East North Central region would be accentuated by the severity of the 1979-82 recession in that region.

The data show that metalworking machinery employment is concentrated in the East North Central, Mid-Atlantic and New England regions. Electronic components employment is concentrated in the same three regions plus the Pacific region, and motor vehicles employment is concentrated in the East North Central region. All three industries show a dispersal of employment from the East North Central and Mid-Atlantic regions to the southern and western regions, with two exceptions. The Pacific states' share of motor vehicle employment declined, and the Mid-Atlantic states' share of motor vehicle employment increased. New England's share of national employment in the three industries remained constant over time.

Table 3.1
Regional Distribution of National Metalworking Machinery, Electronic Component, and Motor Vehicles Employment, 1973-79
(Percent)

	Metalworking machinery		Electronic components		Motor vehicles	
	1973	1979	1973	1979	1973	1979
New England	15.1	15.0	11.1	11.1	1.4	1.4
Mid-Atlantic	19.6	16.2	30.1	20.5	10.3	11.4
East North Central	48.1	46.6	19.7	15.9	66.9	60.8
West North Central	3.6	4.8	4.5	4.0	4.5	7.2
South Atlantic	4.5	5.1	4.9	7.3	5.1	5.2
East South Central	2.1	2.7	2.1	2.7	3.0	4.4
West South Central	1.6	2.6	6.6	6.6	2.7	3.8
Mountain	.5	.9	2.0	4.8	.3	.4
Pacific	5.0	6.1	19.0	27.2	5.8	5.3
Total	100	100	100	100	100	100

SOURCE: The Dun and Bradstreet DMI file, 1973 and 1979.

New England: Maine, Vermont, New Hampshire, Massachusetts, Connecticut, and Rhode Island.
Mid-Atlantic: New York, New Jersey, and Pennsylvania.
East North Central: Ohio, Illinois, Indiana, Michigan, and Wisconsin.
West North Central: North Dakota, South Dakota, Minnesota, Iowa, Missouri, Nebraska, and Kansas.
South Atlantic: Delaware, Maryland, District of Columbia, West Virginia, Virginia, North Carolina, South Carolina, Georgia, and Florida.
East South Central: Alabama, Tennessee, Kentucky, and Mississippi.
West South Central: Texas, Louisiana, Arkansas, and Oklahoma.
Mountain: Montana, Wyoming, Idaho, Utah, Nevada, Colorado, Arizona, and New Mexico.
Pacific: California, Oregon, and Washington.

Table 3.2 shows the distribution of employment across central cities, suburbs and nonmetropolitan areas. Both metalworking machinery and motor vehicles establishments are primarily concentrated in central cities, and electronic components employment is primarily located in suburbs. Both metalworking machinery and motor vehicles employment have become more decentralized over the period, whereas electronic components employment is becoming more centralized.

Table 3.2
Percent of Employment in Central Cities, Suburbs
and Nonmetropolitan Areas, 1973 and 1979

	Metalworking machinery		Electronic components		Motor vehicles	
	1973	1979	1973	1979	1973	1979
Central cities	44.9	40.2	34.8	35.3	52.3	46.5
Suburbs	37.7	42.7	51.1	53.5	33.0	35.1
Nonmetropolitan areas	16.9	16.9	14.1	11.0	14.7	18.5
Total	100	100	100	100	100	100

SOURCE: Dun and Bradstreet DMI file, 1973 and 1979.

Growth Rates by Components of Growth

One of the major strengths of the Dun and Bradstreet data is that they can be used to disaggregate employment growth into the components of change, including employment gains or losses due to (1) establishment start-ups, (2) closings, (3) expansions and contractions, or (4) relocations. Table 3.3 subdivides employment growth rates for the regions by these four components of change and ranks the regions by their employment growth rates, from the slowest- to fastest-growing regions. The data in table 3.3 are graphed in figures 3.1 through 3.3, again with the regions ranked by growth rate from the slowest- to fastest-growing regions. Four generalizations may be derived from these figures.

First, there is a positive association between regional growth rates and employment growth due to start-ups and net expansions. In general, the regions experiencing rapid metalworking machinery and electronic components employment growth appear to be doing well primarily because of the high rate of new business formations in those regions. The regions experiencing high rates of motor vehicles employment growth appear to be doing well because of their high rates of expansions. Second, there is no evidence of a relationship between employment loss due to closings and regional growth rates. In other words, the slow-growth regions do not appear to be growing slowly, or declining, because of high rates of plant failures or shutdowns. Third, the fast-growth regions tend to experience greater employment gains due to net expansions than the slow-growth regions; and fourth, while the migration pattern behaves as expected, with the northern regions experiencing net outmigration and the South and West experiencing net inmigration, these shifts do not play an important role in regional growth and decline.

These conclusions are supported by the results of three simple regression equations, where employment growth is regressed against components of growth. The equations are as follows:

1) $G_r = \alpha + \beta_1 C_r + \beta_2 DM + \beta_3 DE$
2) $G_r = \alpha + \beta_1 S_r + \beta_2 DM + \beta_3 DE$
3) $G_r = \alpha + \beta_1 E_r + \beta_2 DM + \beta_3 DE$

where: G=annual average rate of employment growth.

C=annual average rate of job loss due to plant closures.

S=annual average rate of job gains due to plant start-ups.

E=annual average rate of job gains (or losses) due to net expansions or contractions.

DM=dummy variable for metalworking machinery.

DE =dummy variable for electronic components.

The β_1 coefficients and their respective t-statistics are reported in table 3.4. The results reported in table 3.4 further support the argument that regional variations in industry growth are the result of variations in job creation due to plant start-ups and net expansions and not the consequence of regional variations in plant closures.

Table 3.3
Regional Growth Rates by Component of Employment Change, 1973-79
Annual Average Percent
Regions are Ranked by Industry Growth Rate from Slowest to Fastest Growth

Metalworking Machinery

	Mid-Atlantic	New England	East North Central	West North Central	Pacific	South Atlantic	East South Central	West South Central	Mountain	All U.S.
Start-ups	1.6	1.4	2.6	3.7	4.7	4.3	4.2	10.6	10.3	2.7
Closings	-3.0	-2.3	-3.0	-3.3	-4.4	-2.9	-2.5	-4.6	-3.5	-3.0
Net Expansion	-0.8	0.4	0.7	1.5	1.7	1.1	1.9	2.7	2.6	.5
Net Migration	0.3	-0.5	-0.1	0.1	0.6	0.3	0.1	.0	0.1	.0
	-1.9	-1.0	0.2	2.0	2.6	2.8	3.7	8.7	9.5	.2

Electronic Components

	Mid-Atlantic	West North Central	East North Central	West South Central	New England	South Atlantic	Pacific	Mountain	East South Central	All U.S.
Start-ups	2.4	3.0	4.1	3.6	4.8	6.2	5.9	8.9	17.5	4.5
Closings	-4.2	-4.0	-3.2	-4.2	-3.3	-4.3	-4.5	-2.8	-1.6	-3.9
Net Expansions	-0.8	0.6	0.6	2.9	1.2	1.8	2.5	3.7	1.7	1.1
Net Migration	.0	.0	-0.3	0.2	0.1	0.1	0.1	-0.2	0.1	.0
	-2.6	-0.4	1.2	2.5	2.8	3.8	4.0	9.6	17.7	1.7

Motor Vehicles

	Pacific	Mountain	East North Central	Mid-Atlantic	South Atlantic	New England	West South Central	West North Central	East South Central	All U.S.
Start-ups	0.5	1.6	0.1	0.2	0.3	0.5	1.0	0.3	0.3	.2
Closings	-4.2	-4.3	-2.4	-2.1	-2.4	-2.2	-3.4	-1.8	-1.5	-2.4
Net Expansion	0.8	-0.1	-0.5	-0.3	0.5	0.2	1.8	2.2	2.5	-.1
Net Migration	.0	0.0	.0	.0	0.1	0.4	0.1	0.1	-0.1	.0
	-2.9	-2.8	-2.8	-2.2	-1.5	-1.1	-0.5	0.8	1.2	-2.3

SOURCE: Dun and Bradstreet Data, 1973 and 1979.

Regional Growth by Components of Growth, 1973–79
Ranked from Region with Slowest to Fastest Industry Growth

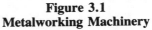

Figure 3.1
Metalworking Machinery

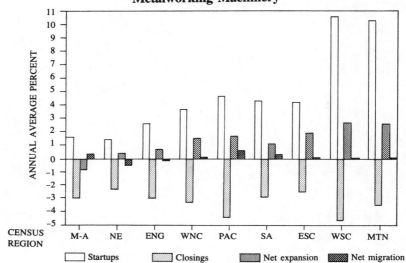

Figure 3.2
Electronic Components

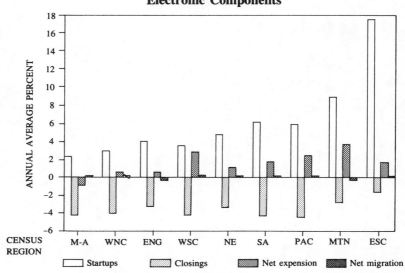

Regional Growth by Components of Growth, 1973–79
Ranked from Region with Slowest to Fastest Industry Growth

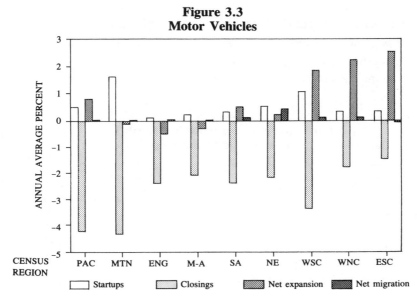

Figure 3.3
Motor Vehicles

Startups   Closings   Net expansion   Net migration

Finally, although these general patterns hold, there are exceptions. For example, the Pacific region experienced higher rates of metalworking machinery employment growth due to start-ups than the East South Central and South Atlantic regions, yet the Pacific region experienced slower aggregate growth due to this region's higher rates of employment losses in plant closings (see table 3.3).

Table 3.4
Results of Regressions, Regional Growth on the
Components Growth, 1975-1982

	Closings	Start-ups	Net expansions
Coefficient	1.53	1.20*	2.27*
t-statistics	(1.70)	(15.00)	(3.98)

SOURCE: From data reported in table 3.3. Calculated from the Dun and Bradstreet Data.
* Statistically significant at the .01 percent level of confidence.

The data are partitioned in a similar fashion for central cities, suburbs, and nonmetropolitan areas in table 3.5. Figures 3.4 through 3.6 show the components of growth, again ranked from the slowest- to fastest-growth areas. The data indicate that metalworking machinery employment is growing fastest in the suburbs, electronic components employment is growing fastest in central cities, and motor vehicle employment is growing fastest in nonmetropolitan areas.

Table 3.5
Annual Average Employment Growth Rates
in Central Cities, Suburbs, and Nonmetropolitan Areas, 1973-79

	Metalworking machinery		
	Central City	Nonmetropolitan	Suburbs
Start-ups	4.3	4.9	7.1
Closings	−7.0	−5.7	−5.4
Net expansions	0.5	2.9	1.9
Migration	−0.2	−0.2	0.4
Total	−2.4	−1.9	4.0

	Electronic components		
	Nonmetropolitan	Suburbs	Central City
Start-ups	6.3	8.9	10.4
Closings	−8.3	−8.1	−7.3
Net expansions	2.1	2.3	3.5
Migration	−0.9	0.4	−0.9
Total	−0.8	3.5	5.7

	Motor vehicles		
	Central City	Suburbs	Nonmetropolitan
Start-ups	0.3	0.5	0.6
Closings	−4.8	−5.2	−4.1
Net expansions	−0.9	0.7	2.7
Migration	−0.2	0.3	0.0
Total	−5.6	−3.7	−0.8

SOURCE: Dun and Bradstreet DMI file, 1973-79.

Intraregional Growth by Components of Growth, 1973-79
Ranked from Region with Slowest to Fastest Industry Growth
(Annual average percent)

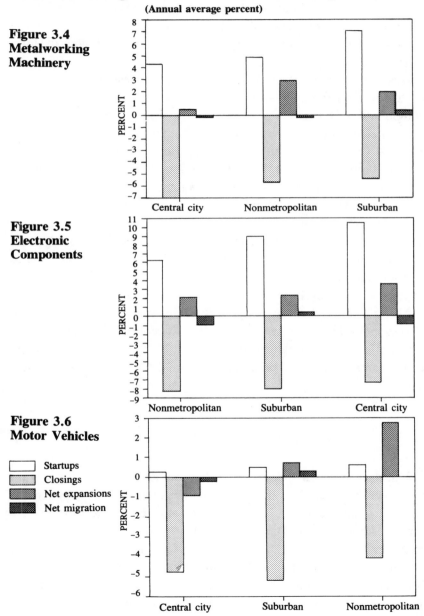

Figure 3.4
Metalworking
Machinery

Figure 3.5
Electronic
Components

Figure 3.6
Motor Vehicles

□ Startups
▨ Closings
▨ Net expansions
■ Net migration

For all three industries the intraregional pattern is similar to the regional pattern in that high industry growth rates are explained by high start-up rates and, for motor vehicles, by expansions rather than by low rates of employment losses due to plant closures. Again, there is no evidence that rates of job loss due to plant closures are higher in slow-growth than fast-growth regions.

David Birch used the total D&B file, including establishments from all sectors, to examine the regional pattern of employment growth by components of change. Birch's work covers the time period 1969 to 1976, and he disaggregates the data into the four major regions of North East, North Central, South, and West. Birch reports three findings similar to ours. One, that the rate of job loss through closures and on-site contractions is fairly constant across regions, approximately 8 percent per year between 1969 and 1976. Two, that the difference in growth rates between fast- and slow-growing regions appears to lie in the rate of job creation due to expansions and births of new firms. And three, that interregional migration accounts for an insignificant percentage of regional employment change (Birch 1979). In spite of different adjustments to the basic D&B data base (some of which are described in chapter 2) and their coverage of different years, these findings are also supported by the Brookings studies (Armington and Odle 1982a). The similarity in our results indicates that Birch's and the Brookings Institution's results hold when the data are disaggregated by industry and that our findings can be generalized to most industries.

Regional Rates of Job Loss

This section looks specifically at how important plant closings are in total employment loss and estimates rates of total job loss by region. Table 3.6 and figures 3.7 through 3.9 show rates of job loss by the nine census regions. Job loss is defined here as those employment losses by either a plant closing, the contraction of a continuing establishment, or the relocation of a plant out-of-state. Job loss may approximate, but is not equivalent to worker displacement. First, when a plant is relocated some employees move with the enterprise. Second, workers laid off an on-going business may be rehired. Third, a job may be lost but no displacement occur because of attrition.

To minimize the inclusion of temporary layoffs in the number of jobs lost due to establishment contractions, the estimates in table 3.6 include only the expansionary period, 1975-79. Workers laid off during a recession, such as occurred from 1973 to 1975 or 1979 to 1982, are more likely to be cyclically unemployed than workers laid off during an economic expansion.

The regions in figures 3.7 through 3.9 are again ranked by industry growth rates. In all industries, plant closings are responsible for the majority of lost jobs. A relatively small proportion of jobs appear to be lost due to plant outmigrations.

For the metalworking machinery and electronic components industries, rates of job loss are no higher in the slow-growth than the fast-growth regions. This conclusion would hold even if the definition of displacement were limited to plant closures and excluded relocations and contractions.

The pattern for motor vehicles is different. As shown in figure 3.9, jobs lost due to plant closures, contractions, and relocations are greater when the regional growth is slower. Contrary to the findings reported in table 3.3 and figure 3.3 from the years 1973 to 1979, this association is due in large part to the higher rates of job loss due to plant closures in the slow-growth regions. The motor vehicle industry may behave differently from metalworking machinery and electronic components for the following reason. When industry employment shrinks in a region, it may first contract by the failure of job gains from start-ups to replace job losses due to an average rate of closings. As the industry declines even further, and the start-up rate is already very low, further industrial decline must occur through rising closure rates.

The 1975-79 motor vehicle results may differ from those of 1973-79 because start-ups may have originally fallen during the 1973-75 recession. As this industry continued to decline during the 1975-79 national recovery, there were few industry start-ups to absorb further economic decline.

To summarize, in the years 1975 to 1979, the rate of motor vehicle job loss was greater in the regions where motor vehicle employment was declining, but this pattern did not hold for the metalworking

Table 3.6
Employment Losses by Reason, 1975-79
Annual Average Percent
Regions are Ranked by Industry Growth Rate from Slowest to Fastest Growth

Metalworking Machinery

	Mid-Atlantic	New England	East North Central	West North Central	Pacific	South Atlantic	East South Central	West South Central	Mountain	All U.S.
Closings	-5.2	-2.6	-5.2	-4.4	-7.0	-4.5	-3.5	-7.0	-9.3	-4.9
Contractions	-4.7	-2.7	-2.4	-1.8	-1.4	-1.0	-0.8	-1.2	-0.8	-2.7
Relocations*	-0.1	-1.5	-0.3	.0	-0.1	.0	-0.4	-0.1	-0.1	-.4
	-10.0	-6.8	-7.9	-6.2	-8.5	-5.5	-4.7	-8.3	-10.2	-8.0

Electronic Components

	Mid-Atlantic	West North Central	East North Central	West South Central	New England	South Atlantic	Pacific	Mountain	East South Central	All U.S.
Closings	-4.3	-6.8	-6.0	-10.3	-6.0	-8.2	-8.6	-4.0	-1.5	-6.6
Contractions	-5.0	-2.6	-2.4	-4.6	-2.7	-0.9	-3.0	-2.4	-1.0	-3.4
Relocations*	-0.1	-0.1	-0.7	.0	-0.1	-0.3	.0	-0.8	.0	-.2
	-9.4	-9.5	-9.1	-14.9	-8.8	-9.4	-11.6	-7.2	-2.5	-10.2

Motor Vehicles

	Pacific	Mountain	East North Central	Mid-Atlantic	South Atlantic	New England	West South Central	West North Central	East South Central	All U.S.
Closings	-7.9	-5.1	-4.9	-3.3	-3.9	-2.6	-5.0	-3.2	-2.1	-4.6
Contractions	-1.8	-2.8	-4.5	-4.6	-1.0	-4.4	-1.2	-2.0	-1.4	-3.8
Relocations*	-0.1	.0	.0	-0.1	-0.2	.0	.0	.0	-0.2	.0
	-9.8	-7.9	-9.4	-8.0	-5.1	-7.0	-6.2	-5.2	-3.7	-8.4

SOURCE: Dun and Bradstreet DMI file, 1975 and 1979.

*Plant out-migration only.

Employment Losses by Reasons, 1975–79
Ranked from Region with Slowest to Fastest Industry Growth
(Annual average percent)

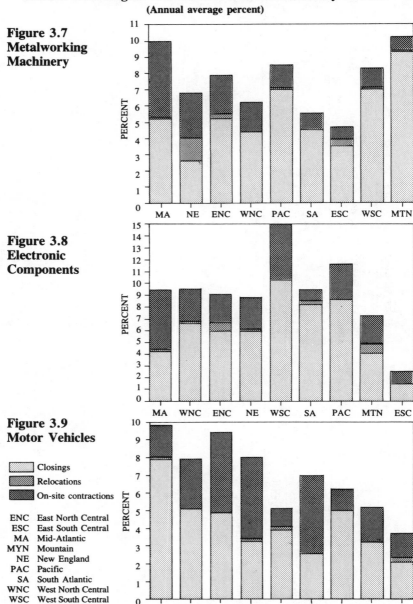

Figure 3.7
Metalworking
Machinery

Figure 3.8
Electronic
Components

Figure 3.9
Motor Vehicles

Closings
Relocations
On-site contractions

ENC   East North Central
ESC   East South Central
MA   Mid-Atlantic
MYN   Mountain
NE   New England
PAC   Pacific
SA   South Atlantic
WNC   West North Central
WSC   West South Central

machinery and electronic components industries, where there is no evidence of a correlation between economic decline and job loss.

Numbers of Jobs Lost By Regions

Another issue relevant to state and federal policymakers concerned with plant closing legislation is where displaced workers are concentrated. The numbers of displaced workers are highest in the regions with the largest employment base (see table 3.7). For example, the East North Central region, with the largest share of national metalworking machinery employment, lost nearly 80,000 jobs between 1975 and 1979, whereas the Mountain region, the area with the lowest percent of national metalworking machinery employment, lost only an estimated 1,542 jobs during this period. The region experiencing the largest losses in electronic component employment was the Pacific region (87,000 jobs), and while the largest motor vehicle employment losses were in the East North Central region (299,107 jobs).

Regional Variations in Plant Closing Rates
and the Structure of Local Economies

Surprisingly, there is no evidence of a negative relationship between rates of job loss (due to plant closings, relocations, or contractions) and economic growth in the metalworking machinery and electronic components industries. There is, however, some evidence that, during the 1975-79 expansion, rates of worker displacement in the motor vehicle industry were highest in the regions where motor vehicle employment growth was slowest.

In this section, we test the hypothesis that regional variations in plant closure rates exist, but are hidden by differences in the age, size and status composition of establishments in the regional economy. For example, headquarters of a given size and age may close at lower rates in the Sun Belt than the Frost Belt, but the Sun Belt regions may still have a high overall closure rate. The reason is that a large proportion of Sun Belt employment could be in new, small plants and in branch

Table 3.7
Number of Jobs Lost, by Region, 1975-79

	New England	Mid-Atlantic	East North Central	West North Central	South Atlantic	East South Central	West South Central	Mountain	Pacific	All U.S.
Metalworking Machinery										
Closings	8155	21741	52763	3229	4020	1536	2059	1415	7207	102125
Relocation	4578	598	2705	25	0	26	20	0	50	8002
Contractions	8635	19671	24256	1305	872	369	341	127	1433	57009
Total	21368	42010	79724	4559	4892	1931	2420	1542	8690	167136
Electronic Components										
Closings	20557	32094	31302	9184	11978	808	24531	2541	64425	197420
Relocation	203	1066	3660	79	480	5	22	490	23	6028
Contractions	9412	37439	12326	3523	1327	520	1101	1564	22579	89791
Total	30172	70599	47288	12786	13785	1333	25654	4595	87027	293239
Motor Vehicles										
Closings	1719	15777	155223	6729	8596	3170	5604	756	21123	218697
Relocation	23	396	745	17	351	250	12	5	275	2074
Contractions	2908	22137	143139	4340	2103	2011	1336	414	4747	183135
Total	4650	38310	299107	11086	11050	5431	6952	1175	26145	403906

SOURCE: Dun and Bradstreet DMI file, 1975 and 1979.

plants and subsidiaries that close at higher than average rates. Here we test whether the closure rates of similar types of plants vary across the regions.

Closing rates are expected to be particularly high for new firms, small establishments, and branch plants and subsidiaries. Opening a new business is a risky venture. The perceived market may not exist or may not be large enough to sustain the new enterprise. Another common reason new firms fail is that many entrepreneurs do not have sufficient capital to sustain the company through several years of losses. Banks are less likely to lend money to companies without a proven track record, and new companies are generally too small for other sources of capital, such as public capital markets, private stock sales, or bond placements. Thus, regional economies with a high proportion of employment in new firms should, all other factors held constant, experience relatively high closure rates.

Small companies also experience relatively high closing rates. Small firms have limited access to financial capital and are therefore more likely to be forced out of business than larger firms when personal sources of revenue or retained earnings run short (Litvak and Daniels 1979). In addition, during bad economic times larger establishments can contract, whereas smaller establishments close. Thus, regions with a high proportion of employment in small establishments may also experience higher rates of plant closings.

A number of authors have speculated that branch plants and subsidiaries provide a less stable employment base than single-plant enterprises. One theory is that managers of multiplant firms allocate capital more efficiently and rationally than managers of single-plant firms and therefore are quick to close branch plants that do not yield sufficiently high rates of return. A second theory argues that large corporations are more likely to mismanage acquisitions, resulting in losses and, ultimately, in closings.

Disinvestment and plant shutdowns may occur more readily in multiplant than single-plant firms because managers of the former have a wider scan of investment possibilities. First, the managers of multiplant firms can easily shift cash flows among branches and subsidiaries. Williamson (1975, pp. 147-148) argues that one of the greatest advantages of multidivisional firms is that capital allocation is more efficient internally than in capital markets external to the firm. A multidivisional firm has good information about all of its branches and divisions and therefore can assign cash flows to high yield uses, phasing out, divesting itself of, or closing less profitable operations. In addition, managers of multiplant and multidivisional operations have the expertise and staff to carry out a wide-ranging search outside the firm for more profitable sites or ventures, again phasing out less profitable activities.

The single-plant manager faces a more limited range of alternative investments because the cost of identifying and taking advantage of new investment opportunities is high. Single-plant firms probably do not have the staff to search out new ventures, and they are less likely to have the skills to operate a new line of business even if the opportunity were identified. Thus many small business people will probably continue to invest in and operate ventures that a multiplant firm would shut down (Cyert and March 1963; Hymer 1972, p. 120; Storper and Walker 1984).

In addition, small entrepreneurs will be more likely to keep a plant operating when its profits fall below those of alternative investments because they draw a salary from the business and live in, derive status from, and have a commitment to the community where the business is located. Noneconomic benefits, therefore, may compensate the small entrepreneur for lower rates of return (Westaway 1974, p. 151; Dicken 1976, pp. 404-405; Erickson 1976; McGranahan 1982).

Evidence of plant closures in Philadelphia supports the view that branch plants or multibranch and multidivisional firms demand higher rates of return to operate than do single-plant enterprises. Hochner and Zibman (1982, pp. 204-205) found that over a 10-year period, independent firms did not close until their sales revenues failed to keep pace with inflation. National corporations closed firms that had average sales increases which kept pace with inflation, and multinational corporations closed branches that experienced average sales growth rates surpassing the inflation rate.

Another school of thought explaining why subsidiaries and branches are more susceptible to shutdown than single-plant operations contends that plant disinvestments and shutdowns occur where rates of return are low relative to alternative investment possibilities. In this model, private firms are often responsible for the low return of their branches and subsidiaries. In the most recent merger movement, companies are spending human and financial resources to acquire and divest themselves of existing businesses in order to improve the "bottom line" of their balance sheets. This "paper entrepreneurism" diverts resources from research and development on new product lines and production methods, from the development of new, more efficient management techniques, and from investments in state-of-the-art technologies. This short-run approach to improving profit performance has led to falling rates of profits in many branches and subsidiaries, resulting, in many cases, in their closing (Reich 1983).

In addition, corporations frequently acquire businesses in which they have little operating experience. As a result, a once profitable acquisition may be poorly managed, become unprofitable, and ultimately be closed (Bluestone and Harrison 1982, pp. 151-159). For example, profits in the New England Provision Company, a meat packing plant, began to decline shortly after its acquisition by LTV. It was estimated that management fees, inept management hired by LTV, and required purchasing from other holdings of LTV cost the plant an additional $325,000. Consequently, the New England Provision Company was eventually closed. In another case, Clinton Colonial Press was acquired by Sheller Globe in 1974 during the purchase of Clinton's parent company, an auto parts business. Lacking interest and management expertise in the printing business, Sheller Globe closed the plant in 1976 (C and R Associates 1978, p. 49).

Alternatively, McKenzie (1982) predicts that branches and subsidiaries are *less* likely to close than single-plant operations. He argues that conglomerate ownership should lead to the continued operation of many plants that would close under local ownership. A locally-owned firm, with limited access to capital and without a cushion of profits from other product lines on a different business cycle, might be forced to close a temporarily unprofitable plant for lack of funds to carry it through a financial storm. A conglomerate would spread the losses over a number of varied ventures enabling it to subsidize temporarily unprofitable plants and adding to job stability.

The evidence on the instability of branch plant employment is mixed. Evidence from Great Britain shows that branches are not particularly susceptible to closure (Townroe 1975). Erickson's (1980) findings are similar. He studied branch plant closings in rural Wisconsin and found that the annual closure rate among branches between 1959 and 1977 was 3.3 percent. This figure compares favorably with an average annual closure rate for all firms in the U.S. economy of 8 percent.

Other authors have found contradictory results. Barkley (1978) found that, over the period 1965-75, branch plants in rural Iowa were more likely to close than locally-owned plants, and Bluestone and Harrison found that corporations and conglomerates together were reponsible for a disproportionate share of job loss through plant closings in New England (1982, p. 34). Barkley's and Bluestone and Harrison's findings are consistent with the findings presented here.

The distribution of new firms, small establishments, and branch plants and subsidiaries should vary across local economies for several reasons. First, as shown above, fast-growth economies have relatively high plant start-up rates. Thus, fast-growth economies should have a relatively high percent of employment in new firms. Second, Armington and Odle (1982b) found small firms to be larger contributors to growth in the slow-growth regions. They present, as one explanation, the fact that large business expansion plans may dominate the short-run supply of labor, locations, and capital and squeeze out small firms in the fast-growth regions. In the less competitive economies, small firms fill the gap left by the retreat of the larger companies. Finally, a number of authors have argued that southern economies are more dependent on branch plant employment than northern economies. According to this argument, economic growth in the southern states is due, in large part, to the expansion of northern firms into the southern economy.

Data from the Dun and Bradstreet file indicate that the phenomenon of southern branches controlled by Frost Belt headquarters is significant. Birch (1979, pp. 45-46) found that over the period 1960-1976, Frost Belt firms controlled nearly 70 percent of the South's net job growth in manufacturing. Analyzing data on the computer industry, Hekman (1980, p. 12) found that New England and the Mid-Atlantic states were more likely to be home to headquarters than branches and that the South

Atlantic and South Central regions had few headquarters relative to the branches they contained.

Southern economies may be relatively dependent on branch plant employment for two reasons. First, the product-cycle model implies that the South will have high concentrations of branch plants in mature industries. Routinized production processes and standardized products can be produced where workers are low-skilled and produced more cheaply where wages are low, i.e., the southern states. In addition, firms producing market-oriented products may have also spun off branches to the South to take advantage of expanding markets in this region resulting from above-average population and employment growth. Headquarters of both production-oriented and market-oriented firms are expected to have remained in the metropolitan areas of the Frost Belt and Pacific regions to retain access to financial markets, to benefit from agglomeration economies, and to operate in proximity to a skilled labor pool (Hymer 1972).

To test the extent to which plant closing rates vary across the regions, we estimate a model predicting the probability a plant will close, based on its regional location, and its size, status as an independent, headquarters, branch, or subsidiary, and its age. The model is estimated with a logit procedure. The logit model is appropriate for cases where the dependent variable is binary (0-1). In this case the dependent variable takes on the value of 1 when an establishment closed and 0 when it did not. The parameters are maximum likelihood estimates. Regional locations and establishment status enter the equations as dummy variables. New England is the reference economy for the regional dummy variables. Independent establishment is the reference catagory for the status dummy variables.

Three models are estimated. One includes just the regional dummies and tests the hypothesis that regional location influences the probability a plant will close. The second model includes, in addition to the regional dummies, the size and status of the plant. The third model includes only single-plant firms. This model is estimated to test the impact of age on the probability a plant will close. The age variable could not be included in the equation estimated with all establishments, since dates of birth are not available for branch plants.

The models are estimated with D&B data from the 1975-82 period. This period is one full business cycle, including the 1975 to 1979 expansion and the 1979 to 1982 recession. These data also capture the most recent years in our data set. Data from the more recent cycle is preferable, since the DMI file has become more accurate with time.

The results of the three models are presented in tables 3.8, 3.9 and 3.10, respectively. In all three tables, the regional results are similar, whether or not we control for the status, age, and size of establishment. In table 3.8, the significant coefficients indicate plants in the Sun Belt are more likely to close than plants in the Frost Belt for the metalworking machinery and electronic components industries. In all three industries, there is only one case where the coefficient is significant and of the expected sign. This is in the motor vehicles equation, where the coefficient on East South Central is negative, indicating that the probability of a closure is lower in the region where motor vehicle employment is growing most rapidly.

Table 3.9 presents the results when plant status and size are held constant. The results indicate that, as hypothesized, larger establishments are less likely to close than smaller establishments. The coefficients for this variable are statistically significant at very high levels of confidence. Headquarters appear to close at about the same rate as independents in the metalworking machinery industry, but at lower rates than independents in the electronic components and motor vehicles industries. Both branches and subsidiaries are more likely to close than independents. The coefficients on these variables are not only large relative to all other coefficients, but are significant at very high levels of confidence.

Table 3.10 reports the results for single-plant firms (independents) only. Older independents are more likely to close in the metalworking machinery and electronic components industries, but not in the motor vehicle industry. Size is statistically significant in all three industries, with large firms less likely to close than small firms. Again, when the regional variables are statistically siignificant they show higher plant closer rates in the faster-growth Sun Belt economies.

We estimated the same equations for each industry for the periods 1975 to 1979 and 1979 to 1982. Breaking the data into recession and expansion does not alter the general conclusion that closing rates tend to be slightly higher in the faster- than in the slower-growth economies. We also estimated the equation with dummies for central cities, suburbs, and nonmetropolitan areas in each region. This was done to determine the extent to which regional aggregates were masking variations in plant closing rates within regions. These results are reported in appendix 3.1 and do not alter the conclusions reached here.

Table 3.8
Logit Estimates of the Probability of a Plant Closing
by Region, 1975-82

Independent variables	Metalworking machinery	Electronic components	Motor vehicles
Constant	−.49*	−.02	−.11
	(99.03)	(.12)	(.86)
Mid-Atlantic	.03	−.15**	.02
	(.26)	(2.97)	(.03)
East North Central	−.11*	.01	−.05
	(4.13)	(.03)	(.17)
West North Central	.01	−.002	−.08
	(.00)	(.00)	(.31)
South Atlantic	.31*	.14	−.02
	(10.72)	(1.36)	(.01)
East South Central	.10	.18	−.40*
	(.78)	(.65)	(5.40)
West South Central	.57*	.27*	.04
	(26.45)	(4.42)	(.09)
Mountain	.56*	.09	.01
	(14.32)	(.34)	(.00)
Pacific	.32*	.08	.13
	(19.92)	(1.01)	(1.03)
No. of observations	14745	6372	6402
Fraction of pairs where predicted = actual response	.40	.40	.45

Figures in ( ) are Chi-square statistics.
*Significant at the 5 percent level of confidence or greater.
**Significant at the 5 to 10 percent level of confidence.

The apparent inconsistency, in the motor vehicle industry, between even closure rates across the regions and uneven rates of job loss due

Table 3.9
Logit Estimates of Plant Closings
by Region and Status, 1975-82

Independent variables	Metalworking machinery	Electronic components	Motor vehicles
Constant	−.55*	−.14*	−.11
	(120.32)	(3.79)	(.83)
Size	−.001*	−.0004*	−.0003*
	(23.69)	(20.45)	(38.17)
Headquarters	.10	−.16*	−.33*
	(1.77)	(3.53)	(14.10)
Branch	1.00*	.83*	.47*
	(228.18)	(126.70)	(42.58)
Subsidiary	.68*	.43*	.29*
	(52.69)	(20.72)	(7.03)
Mid-Atlantic	−.02	−.14	.02
	(.14)	(2.42)	(.01)
East North Central	−.14*	.01	−.08
	(5.99)	(.02)	(.38)
West North Central	−.01	−.02	.05
	(.01)	(.01)	(.11)
South Atlantic	.25*	.07	−.08
	(6.60)	(.38)	(.32)
East South Central	−.04	.08	−.50*
	(.12)	(.13)	(8.21)
West South Central	.52*	.27*	.005
	(21.29)	(4.31)	(.00)
Mountain	.53*	.07	−.03
	(12.45)	(.19)	(.02)
Pacific	.31*	−.16	.11
	(17.85)	(1.11)	(.67)
No. of observations	14745	6372	5811
Fraction of pairs where predicted=actual response	.52	.54	.54

Figures in ( ) are Chi-square statistics.
*Significant at the 5 percent level of confidence or greater.

to plant closures across the regions may be explained by regional differences in the average size of the closing establishment, with larger plants closing in the slow-growth regions. The data in table 3.11 do not bear this out, however. The average size of the closing plant is not negatively associated with economic growth. The number of workers in the average closing plant is substantially larger in the East North Central than in any other region, and this may explain why we found a relatively large rate of job loss in the East North Central (table 3.6) with an average rate of plant closures (table 3.3).

Table 3.10
Logit Estimates for Probability of a Closing,
Single-Plant Firms Only, 1975-82

Independent variables	Metalworking machinery	Electronic components	Motor vehicles
Constant	−.42*	−.11	−.08
	(55.32)	(1.49)	(.39)
Age	.001*	.001*	.000
	(6.33)	(6.85)	(.02)
Size	−.009*	−.0007**	−.005*
	(85.72)	(2.81)	(21.58)
Mid-Atlantic	−.05	−.20**	.002
	(.43)	(3.27)	(.00)
East North Central	−.20*	−.09	−.06
	(10.30)	(.61)	(.17)
West North Central	−.05	−.07	.20
	(.19)	(.14)	(1.18)
South Atlantic	.22*	.22	−.06
	(4.08)	(2.23)	(.13)
East South Central	.16	.06	.02
	(1.38)	(.04)	(.01)
West South Central	.49*	.18	.06
	(15.62)	(1.28)	(.14)
Mountain	.46*	.17	.06
	(8.35)	(.88)	(.07)
Pacific	.24*	−.04	.06
	(8.84)	(.16)	(.15)
No. of observations	12116	4032	3483
Fraction of pairs where predicted=actual response	.56	.48	.52

Figures in ( ) are Chi-square statistics.
*Statistically significant at the 5 percent level of confidence or greater.
**Statistically significant at the 5 to 10 percent level of confidence.

Table 3.11
Average Number of Employees in Closing Plants in Motor Vehicles Ranked by Region with Slowest to Fastest Employment Growth

Region								
PAC	MTN	ENC	MA	SA	NE	WSC	WNC	ESC
60	12	332	77	80	31	33	47	77

SOURCE: Dun and Bradstreet data, 1975, 1979, and 1982.

One explanation for higher closing rates in the Sun Belt regions may be a bias in the data. As indicated in chapter 2, D&B drops many continuing independent firms from the file. These tend to be firms that moved; frequently, such enterprises are growing and in need of additional space (Schmenner 1982). Since the data suggest that growing and expanding firms are more likely to be in the Sun Belt, it is possible that D&B dropped more continuing firms in this region than in the slow-growth regions. If this is the case, the plant closure rates in the Sun Belt would tend to be overstated.

Summary

Three conclusions are drawn from the above data. First, there is little regional variation in plant closure rates, and, when there are differences, the rates are higher in the faster-growth Sun Belt economies than in the Frost Belt.

Second, in the metalworking machinery and electronic components industries, there is no evidence that a worker is more likely to be displaced in areas where the industry is slow-growing than growing rapidly. Conversely there is no evidence that high rates of job loss explain regional industrial decline. Instead, rates of displacement are much the same in all regions and area variations in employment growth are primarily due to variations in start-up rates and expansions. There is some evidence, however, that rates of displacement from the motor vehicle industry are higher in the regions experiencing the greatest losses in motor vehicle employment. It is possible that in industries such as motor vehicles where births are already very low, further economic decline must occur through increased rates of plant closurees.

Third, we find that most of the jobs lost in a region, are lost through plant closings. The second most important cause of job loss is contractions, while relocations play only a small role. These conclusions argue that a closure is no more likely to occur in a declining than a growing economy, not that the consequences of a plant closure are equally severe in all regions, a topic explored in more detail in chapter 5. In chapter 4, we directly address the causes of plant closures.

Appendix 3.1

Logit Estimates for Probability of a Closing
by Central City, Suburb, and Nonmetropolitan Area
by Region, 1975-1982

Independent variables	Metalworking machinery	Electronic components	Motor vehicles
Constant	−.63*	−.07	−.08
	(45.53)	(.26)	(.28)
Size	−.001*	.0003*	.000*
	(23.99)	(20.32)	(39.51)
Headquarters	.09	−.16**	−.32*
	(1.66)	(3.35)	(13.02)
Branch	1.02*	.88*	.50*
	(232.58)	(135.85)	(48.48)
Subsidiary	.68*	.44*	.32*
	(52.24)	(21.09)	(8.18)
New England–Sub	.21*	−.04	−.18
	(2.93)	(.05)	(.44)
New England–NM	−.03	−.22	.28
	(.04)	(1.29)	(.77)
Mid-Atlantic–CC	.26*	−.22	.06
	(5.07)	(1.68)	(.07)
Mid-Atlantic–Sub	.10	−.19	−.06
	(.90)	(1.63)	(.10)
Mid-Atlantic–NM	−.28**	−.32**	−.05
	(3.65)	(2.76)	(.04)
E. No. Central–CC	.01	.18	−.12
	(.00)	(1.09)	(.50)
E. No. Central–Sub	−.08	−.12	.11
	(.69)	(.56)	(.37)
E. No. Central–NM	−.13	−.43*	−.40*
	(1.28)	(4.70)	(4.82)
W. No. Central–CC	.18	−.04	.11
	(1.18)	(.01)	(.24)
W. No. Central–Sub	.04	.11	.01
	(.06)	(.17)	(.00)
W. No. Central–NM	−.04	−.42	−.05
	(.04)	(2.24)	(.06)
South Atlantic–CC	.28**	.08	−.39**
	(2.72)	(.12)	(2.98)
South Atlantic–Sub	.36*	.06	.19
	(4.83)	(.10)	(.67)

Appendix 3.1 (continued)

Independent variables	Metalworking machiner	Electronic components	Motor vehicles
South Atlantic–NM	.32**	–.20	–.13
	(3.38)	(.76)	(.35)
E. South Central–CC	–.05	.44	–.53**
	(.07)	(1.52)	(3.61)
E. South Central–Sub	–.18	.25	–.50
	(.43)	(.14)	(1.76)
E. South Central–NM	.35*	–.49	–.55*
	(3.78)	(1.97)	(5.52)
W. South Central–CC	.66*	.33**	.03
	(16.12)	(2.85)	(.03)
W. South Central–Sub	.57*	.11	–.22
	(5.86)	(.17)	(.66)
W. South Central–NM	.50*	–.21	–.02
	(4.70)	(.42)	(.01)
Mountain–CC	.77*	–.10	–.12
	(12.51)	(.19)	(.18)
Mountain–Sub	.35	.30	–.26
	(1.84)	(1.08)	(.48)
Mountain–NM	.78*	–.37	.24
	(3.84)	(.82)	(.47)
Pacific–CC	.48*	.18	.07
	(13.80)	(1.35)	(.14)
Pacific–Sub	.33*	–.09	.09
	(8.16)	(.38)	(.26)
Pacific–NM	.38	–.07	–.001
	(1.72)	(.05)	(.00)
No. of observations	14745	6372	5811
Fraction of pairs where predicted=actual response	.55	.57	.52

*Statistically significant at the 5 percent level of confidence of greater.
**Statistically significant at the 5 to 10 percent level of confidence.
( ) Chi-square statistics.

REFERENCES

Armington, Catherine and Marjorie Odle (1982a) "Sources of Recent Employment Growth, 1978-80," unpublished paper prepared for the second annual Small Business Research Conference, Bentley College, Waltham, Massachusetts, March 11-12.

Armington, Catherine and Marjorie Odle (1982b) "Small Business—How Many Jobs?" *The Brookings Review* (Winter), pp. 14-17.

Barkley, David L. (1978) "Plant Ownership Characteristics and the Location Stability of Rural Iowa Manufacturers," *Land Economics* 54, 1, pp. 92-100.

Birch, David (1979) *The Job Generation Process.* Cambridge, Massachusetts: MIT Program on Neighborhood and Regional Change.

Bluestone, Barry and Bennett Harrison (1982) *The Deindustrialization of America.* New York: Basic Books.

C and R Associates (1978) *Community Costs of Plant Closings: Bibliography and Survey of the Literature.* Report prepared for the Federal Trade Commission, July.

Carlton, Dennis (1979) "Why New Firms Locate Where They Do," in William Wheaton (ed.) *Interregional Movements and Regional Growth.* Washington, D.C.: Urban Institute.

Cyert, Richard M. and James G. March (1963) *A Behavioral Theory of the Firm.* New Jersey: Prentice Hall.

Dicken, Peter (1976) "The Multiplant Business Enterprise and Geographical Space: Some Issues in the Study of External Control and Regional Development," *Regional Studies* 10, pp. 401-412.

Engle, Robert (1974) "A Disequilibrium Model of Regional Investment," *Journal of Regional Science* 14, 3, pp. 367-376.

Erickson, Rodney (1976) "Non Metropolitan Industrial Expansion: Emerging Implications for Rural Development," *Review of Regional Studies* 6, 1, pp. 35-48.

Erickson, Rodney (1980) "Corporate Organization and Manufacturing Branch Plant Closures in Non-metropolitan Areas," *Regional Studies* 14, pp. 491-501.

Harris, Candee (1986) "High Technology Entrepreneurship in Metropolitan Industries" in Edward Bergman (ed.), *Local Economies in Transition: Policy Realities and Development Potentials.* Durham, North Carolina: Duke University Press.

Healey, M.J. (1982) "Plant Closures in Multiplant Enterprises—The Case of a Declining Industrial Sector," *Regional Studies* 16, 1, pp. 37-51.

Hekman, John (1980) "The Future of High Technology Industry in New England: A Case Study of Computers," *New England Economic Review* (January/February), pp. 5-17.

Hochner, Arthur and Daniel M. Zibman (1982) "Capital Flight and Job Loss: A Statistical Analysis," in John C. Raines, Lenora E. Berson, and David Gracie (eds.), *Community and Capital in Conflict, Plant Closings and Job Loss*. Philadelphia, Pennsylvania: Temple University Press, pp. 198-210.

Howland, Marie (1984) "Regional Variations in Cyclical Employment," *Environment and Planning A* 16, pp. 863-877.

Howland, Marie (1985) "The Birth and Intraregional Location of New Firms," *Journal of Planning Education and Research* 4, 3, pp. 148-156.

Howland, Marie and George Peterson (1988) "The Response of City Economies to National Business Cycles," *Journal of Urban Economics* 23, 1, pp. 71-85.

Hymer, Steve (1972) "The Multinational Corporation and the Law of Uneven Development," J.N. Bhagwati (ed.), in *Economics and World Order from the 1970's to the 1990's*. New York: Macmillan, pp. 113-135.

Kieshnick, Michael (1979) *Taxes and Growth*. Washington, D.C.: Council of State Planning Agencies.

Litvak, Lawrence and Belden Daniels (1979) *Innovations in Development Finance*. Washington, D.C.: Council of State Planning Agencies.

Massey, Doreen and Richard Meegan (1982) *The Anatomy of Job Loss*. London: Metheun.

McGranahan, David (1982) "Absentee and Local Ownership of Industry in Northwestern Wisconsin," *Growth and Change* 13, 2, pp. 31-35.

McKenzie, Richard (1982) "The Closing of Profitable Plants: An Area for Government Restrictions," *Backgrounder*. Washington, D.C.: The Heritage Foundation.

Reich, Robert (1983) "The Next American Frontier," *The Atlantic Monthly* (March), pp. 43-58.

Schmenner, Roger (1982) *Making Business Location Decisions*. Englewood Cliffs, New Jersey: Prentice Hall.

Storper, Michael and Richard Walker (1984) "The Spatial Division of Labor: Labor and the Location of Industries" in Larry Sawers and William Tabb (eds.), *Sunbelt/Snowbelt: Urban Development and Regional Restructuring*. New York: Oxford University Press, pp. 3-47.

Townroe, P.M. (1975) "Branch Plants and Regional Development," *Town Planning Review* 46, pp. 47-62.

Westaway, J. (1974) "The Spatial Hierarch of Business Organizations and Its Implications for the British Urban System," *Regional Studies* 8, pp. 145-155.

Williamson, Oliver E. (1975) *Markets and Hierarchies: Analysis and Antitrust Implications*. New York: Free Press.

4

Plant Closures
and
Regional Rates of Profit

Regional employment growth is determined by the rate at which jobs are created through establishment start-ups, expansions, and the inmigration of plants, minus the rate at which jobs are phased out by plant closures, contractions, and the outmigration of establishments. There is a well-developed theoretical and empirical literature on the factors that encourage job creation, including the economic conditions that spawn local jobs and draw firms and branch plants to a location (Stafford 1980; Smith 1981). On the other hand, there has been surprisingly little study of the reverse side of location theory, the factors that influence the decision to permanently close or relocate a plant.

Data presented in chapter 3 show that plant closing rates and rates of worker displacement are no higher in areas where industry is declining than where industry is experiencing strong growth. This finding suggests the factors dampening new start-ups in a region may not be the same ones influencing plant closures.

Chapter 4 explores the causes of plant closings by using a model that estimates the probability a plant will close. We hypothesize that the probability of a plant's closing during some period is a function of the plant's profit level at the beginning of the period and the change in the plant's profits during the period compared to profits a plant could have earned in some other region. We also hypothesize that there are regional variations in profit levels and these are the result of regional differences in production costs and revenues. Plant profits are expected to be lower where production costs are relatively high and markets are declining.

The following five hypotheses are tested. Plant closures are influenced by (1) local production cost levels, (2) the extent of unionization of the labor force, (3) changes in local prodution costs relative to changes in

65

costs in other regions, (4) the degree of industry import competition, and (5) local and national market growth.

The chapter begins with a review of previous literature on the causes of plant closures and the conditions found to influence investment. Because the latter issue has received a great deal of attention in the literature, we focus on studies that have also used the Dun and Bradstreet (D&B) DMI file. Next we present the model and report the results of the empirical tests, which use the D&B data described in chapter 2. The D&B data were supplemented with data from the U.S. Bureau of Labor Statistics, U.S. Census, and U.S. Department of Energy.

The statistical results provide little evidence that the economic variables identified here are responsible for plant closures. The only variable that consistently influences the probability a plant will close is the plant's status as a branch, headquarters, independent, or subsidiary, with branches and subsidiaries closing at much higher rates than headquarters and independents. For independents (single-plant firms), the firm's age and size are also factors in the probability the plant will shut down.

Previous Literature

Most empirical research on the causes of plant closures is based on surveys of managers responsible for closing a plant. For example, Schmenner (1982, pp. 237-239) conducted a survey of 175 closing facilities. He found that 21 percent attributed their closing to high labor costs, 10 percent cited unions, 2 percent cited high energy costs, and 1 percent claimed high taxes were factors. Fifteen percent cited competitive pressures from foreign producers and 58 percent cited domestic competition as additional reasons for their plant closings. Schmenner's findings lend support to the hypothesis that spatial differences in factor costs, unionization rates and market growth should result in spatial differences in plant closure rates. Gainer (1986) surveyed 355 managers who closed plants and found 57 percent of them attributed the closing to high labor costs. The Rhode Island Strategic Development Commission (1983) surveyed all manufacturing firms of 50 employees or more that either closed or experienced major layoffs in Rhode Island between

January 1, 1972 and August 31, 1982 to determine the reasons for these job losses. During this period, 177 firms closed. The commission found that the lack of business acumen, overexpansion, overleveraging or in-abiity of new management to successfully operate a business led to the majority of firm failures. They did not find labor problems or energy costs to be a primary cause of firm failure.

Authors of the Rhode Island study believe their results differ from those of Schmenner because the Rhode Island sample is dominated by small independents, whereas Schmenner surveyed Fortune 500 firms. The few national corporations included in the Rhode Island study also reported high costs among their reasons for closure (Reamer 1987).

Massey and Meegan (1982) examined nearly 60 industries to study the different mechanisms of employment decline. They categorized each industry by its causes of displacement, including: job loss due to technological change; intensification of the production process, i.e., increasing worker productivity without substantial technological change; and rationalization of the production process, i.e., cutbacks in declin-ing industries. They found that firms within each of the three categories, in fact within each industry, exhibit a wide variety of responses to the same external circumstances. For example, firms in a declining industry may close plants, diversify product lines, or seek new markets, and as a consequence, there were no obvious spatial patterns to their employ-ment adjustments. They suggest that more generalizations might be possi-ble if the data were analyzed by other categorical breakdowns, such as the size of firm or the nature of the labor process.

Press accounts of closings also lend support to the view that high wages, energy costs and tax rates can be responsible for plant closures. For example, Duracell Inc. closed a lithium battery manufacturing plant in North Tarrytown, New York, laying off 200 employees. John Bergman, a spokesman for Duracell, said the plant had become un-profitable because of high taxes, along with high fuel costs (*New York Times*, 2/2/84, p. 4:6).

While statistical research on plant closures is rare, a number of em-pirical studies have looked at the reverse question of which factors en-courage the birth and location of new firms and the location of branch plants. A number of these studies have used the D&B data. Carlton

(1979) used the D&B data to study the birth and location decisions of plants in the plastic products, electric transmitting equipment, and the electronic components industries. Modeling the start-up and location decision of independents separately from those of branch plants, Carlton found that wages had the greatest impact on the location of single-plant firms. For every 1 percent decrease in wages, new births increased by 1 percent. Aside from wages, the existence of agglomeration economies, the availability of a skilled labor force, and for two of the three industries, low energy costs also had a positive effect on a Standard Metropolitan Statistical Area's (SMSAs) rate of business start-ups. The results for branch plants were similar, except that a local skilled labor force was less important for branches than for independents.

Harris (1986) conducted a study similar to Carlton's, using data on branch and new firm start-ups in the high technology sector. She found SMSAs with a skilled labor force, low energy costs, and agglomeration economies attracted the greatest number of start-ups. Branch plants were drawn to SMSAs with agglomeration economies and a skilled labor force.

Howland (1985) also tested a model of new firm start-ups using the D&B data. Howland's study differed from those of Carlton and Harris in that she used data on 106 counties in New Jersey, Maryland, Washington, D.C., and Virginia from the metalworking machinery and electronic components industries. Her findings show the greatest number of firm births and employment gains due to births were in the counties with low wages near metropolitan areas. Proximity to a Ph.D.-granting university also influenced the start-up of firms in the electronic components industry.

Rees (1979) found the Southwest had become a spawning ground for innovations and growth industries, and not merely a receiving ground for standardized production processes from the northern states. Using data compiled by the Bureau of Business Research at the University of Texas at Austin in combination with the D&B data, Rees found that 61 percent of new firms in Dallas/Fort Worth were firm births and 39 percent were branch plants. Fifty-seven percent of the branches had headquarters in the local area. Rees attributes this strong local job generation to the growth of agglomeration economies and local markets in the Sun Belt.

In general, this literature concludes that firms locate where they can operate most profitably. Results from the above studies, as well as others (Smith 1981; Stafford 1980; Malecki 1981), show most new firms and branches operate most profitably where wages are relatively low, energy costs are low, agglomeration economies exist, and an appropriately skilled labor force is available. Do these same factors increase the profitability of existing establishments and reduce the probability they will close? This is the question addressed below.

The Causes of Plant Closures—Five Hypotheses

We hypothesize that the probability of a plant permanently closing during some period is a function of the profit level at the beginning of the period and the change in profits during the period. A plant will continue to operate at its current site as long as economic profits are positive and as long as profits in the current location exceed profits that could be earned in alternative locations, minus transaction costs associated with relocating, or as long as

(1) $TR_i - TC_i > 0$ and,

(2) $TR_i - TC_i > TR_o - TC_o - T_i$, where

TR = Total revenue

TC = Long run total costs

T_i = Transactions costs associated with relocating or closing plant i

i   = Plant at the current location

o   = Plant's next most profitable location

When the conditions in (1) or (2) no longer hold, the process of disinvestment begins, and closure can be predicted.

Equation (2) makes explicit that factor prices in other locations (TC_o) and transaction costs associated with closing (T_i), including, the plant's scrap value, the values of tax write-offs, severance payments to workers, and site clean-up costs if the plant were to close, influence closures. For example, a reduction in wages at another site would increase the opportunity costs of operating at the current location (by lowering TC_o) and increase the probability of closure. Higher required severance pay-

ment to workers if the plant should close should increase the transactions cost of closure (T_i) and reduce the likelihood of a shutdown.

There are several reasons to expect profit levels to vary spatially and
to result in variations in plant closure rates. Five hypotheses are addressed below.

Hypothesis 1: Regional Differences in the Price of Inputs

One assumption projecting regional variations in rates of plant closures
is that plants in some regions, particularly the industrialized North, have
higher production costs not compensated for by scale economies or
transport costs. These higher production costs result in lower rates of
plant profit, which in turn lead to higher-than-average rates of plant
closures.

Hulten and Schwab (1983) provide evidence that profit rates vary
spatially and are, in fact, lower in the high-cost regions of the U.S.
They found a 15.9 percent annual average rate of return to capital in
the Sun Belt and a 14.2 percent return in the Frost Belt states for the
period 1951 to 1978.

Plants should close at above average rates in the low-profit regions
as establishments both fail at above average rates and relocate to the
lower cost areas of the country.

A number of authors have argued this hypothesis. Leary (1985, p.
115) states:

> High labor costs, aging infrastructure, and increasingly
> burdensome regulatory restrictions. . . . are creating disad-
> vantageous conditions for firms operating in these [North and
> North Central] regions. It is the 'push' of these higher costs
> in the frostbelt . . . . which is causing the trivial amount of
> outmigration.

Mansfield (1985, p. 244) makes the argument for the textile industry.

> Considerable excess capacity existed in the cotton textile in-
> dustry from about 1924 to 1936. . . . It is also important to
> note that profit rates in cotton textiles were higher in the South
> than the North. Profits in the South averaged 6 percent,
> whereas profits inthe North averaged 1 percent. This was

due to the fact that the prices of many inputs—like labor and raw cotton—were lower in the South.

If we are given a situation of this sort, what would the perfectly competitive model predict? . . . Many firms would have to leave the industry. . . . The industry would tend to become concentrated more heavily in the South, the exit rate being higher in the North than in the South.

Bluestone and Harrison (1982) follow a parallel line of reasoning in their book *The Deindustrialization of America.* They argue that regional disinvestment and plant closures in the Frost Belt states are the consequence of both high production costs in the northern states and intensifying competition in many basic U.S. industries.

Firms in some of the United States' basic industries were willing and able to pay higher prices for inputs when output was concentrated in a few oligopolists and profit levels were sizable in the late 1940s, 1950s and 1960s. In the 1970s, the profit margins eroded. From the period 1963-66 to 1967-70, real rates of return for all nonfinancial corporations in the U.S. fell from 15.5 percent to 12.7 percent. By the 1975-1978 period, rates of return had fallen to 9.7 percent.

This profit squeeze was caused primarily by intensified competition and falling market shares. Firms found their northern operations were especially hard hit because of high production costs in this region, and as companies began to search for ways to restore profit margins, they began phasing out plants located in high cost regions and reinvesting their capital in new or existing plants in the lower cost areas, both domestic and foreign.

For example, during the 1950s and 1960s the oligopolistic structure of the American auto industry permitted wage increases, negotiated by union leaders, to be easily passed on to consumers. When U.S. producers were faced with intensifying competition from lower cost Japanese and German imports in the 1970s, revenues and profit levels fell, and the American auto companies could no longer hold their market share while continuing to pay the high wages and benefits negotiated during better economic times. To cut costs in this new competitive climate, companies began to shift production out of the high-cost North Central region to lower cost regions of the U.S. and abroad, and to outsource, or purchase inputs from abroad that were once produced internally by the firm or from other American suppliers. Outsourcing reduces costs when manufactured inputs can be purchased at lower cost than they can be produced "in house" (Bluestone and Harrison 1982).

Hypothesis #1 applies only to firms facing competition, since firms in concentrated industries can pass at least some of the cost increases on to consumers and therefore may not be driven to seek minimum-cost locations. As indicated by Bluestone and Harrison, it was not until the auto industry faced increased competition that it began to shift investment out of the high-cost regions.

This argument is also unlikely to apply to industrial activity in the earliest stages of the product cycle (see Vernon 1966). When a product is at the innovation stage, the product design is still variable and the production process flexible. Producers, therefore, require proximity to a specialized and skilled labor force, to other manufacturers in order to monitor their discoveries, and to venture capital markets. For innovative activities, nearness to these specialized inputs is likely to compensate for high factor costs.

To summarize, we expect profits to be relatively low in locations where wages, energy costs, and taxes are high, referring back to equations (1) and (2), where TC_i are high. Plant closures are expected to be higher in regions where profit levels are relatively low, especially in more competitive industries and where production has become standardized.

Hypothesis 2: Unionization

In addition to higher costs, the extent to which a local labor force is unionized may also influence plant-level profits and the probability a plant will close. Two arguments are relevant here.

The first argument is that a highly unionized labor force reduces plant profit levels by demanding higher wages, receiving more generous fringe benefits, imposing inflexible work rules on the production site, demanding minimum shop-safety standards, and instituting work stoppages and strikes. Holding all other conditions constant, indirect costs may therefore be higher and profits lower in plants with a unionized labor force. Thus, we conjecture that, faced with increased competition and downward pressure on prices, U.S. producers in the unionized regions are more likely to close and to relocate to nonunion labor markets.

Bluestone and Harrison (1982, pp. 147-148) also argue this hypothesis:

> it now seems to be a fair generalization that, for the most part, the geographic patterning of deindustrialization has to do with the proclivity of companies to shift capital out of older industrial areas to escape having to live with a labor force that . . . had become costly. . . . [C]ompanies are . . . shifting operations . . . into the Sun Belt, non-union peripheries within the North and abroad.

Again, this hypothesis may not apply to firms that face little competition or plants producing products in the earliest stage of the product cycle. Firms in concentrated, less competitive industries are less driven to seek out cost-minimizing sites, since they can pass some of their higher costs on to consumers. Plants producing a state-of-the-art product require a highly skilled labor force, and managers may have to work with unions to fulfill this requirement.

A second argument is that unions will increase the likelihood of a closing among firms adopting new technologies. Firms acquiring new technologies generally undergo changes in skill requirements. Some workers become redundant, and employers would prefer to lay them off. Other workers must be retrained for new tasks. Both of these adjustments are more easily made in a nonunion than in a union plant. In plants where unions have negotiated rigid job categories and made layoffs due to technological change costly and difficult, the adjustment to a new technology can be a slow and contentious process. Employers may, therefore, find it less expensive to close the plant and start over in a new location with a new technology and new labor force than to attempt to increase productivity at an existing plant. We therefore predict a higher rate of plant closures in locations where a higher percentage of workers are unionized.

Hypothesis 3: Changes in Factor Prices

There are three reasons to expect that plants located in areas with greater than the national average increase (or a smaller-than-average decrease) in input prices should experience above average rates of plant closures. First, increases in local production costs should reduce profit levels in local plants and force many marginal plants facing elastic demand curves to close. Referring back to equation (1), TC_i will increase, making the condition for closure more likely. Second, the opportunity cost of operating in the now higher cost region will increase, making the now lower cost sites relatively more profitable. This is shown in equation (2), by an increase in TC_i. Managers will respond by relocating to and reinvesting in the now lower cost regions. Third, when increases in labor costs are responsible for cost increases, the company may install labor saving technologies. When these new technologies cannot be adopted at an existing plant, closure is more likely. Again, this hypothesis is expected to apply to firms in competitive industries, since such firms are less able to absorb losses or pass increased costs on to consumers, and to firms not tied to high-cost regions because of a specially skilled labor force.

To summarize, holding all other variables constant, we expect the loss in profits to be greatest and, therefore, plant closures highest in the locations with the greatest increases in production costs.

Hypothesis 4: Import Penetration

Growing dependence on imports, in particular from Third World countries, is receiving widespread blame for the decline in U.S. manufacturing employment and the spread of industrial plant closures. Three examples are cited here:

> [W]ith import penetration rates reaching 50 percent or over for the last three years, hardly a month goes by without the closing of another shoe plant (Fecteau and Mera 1981, p. 88).

> There are things happening as a result of our foreign trade policies that are undermining U.S. economic strength—and these things should not go uncorrected. These things include the loss of jobs to imports (Finley 1981, p. 78).

> Too many dollars spent by U.S. consumers are stimulating jobs in Asia, rather than Alabama. Troubled U.S. manufac-

turers are shutting U.S. plants or shifting abroad, as cheap imports clobber U.S. companies (Phillips 1984, p. F1).

Both price competition from abroad and the loss of markets are reducing revenues and cutting into the profits of U.S. producers, especially those producers in mature industries utilizing a low-skilled U.S. labor force. We conjecture, therefore, that plant closures are more likely to occur in industries and regional economies dependent on a low-skilled labor force.

The product cycle model, referred to above, predicts changes in an industry's locational requirements as that industry's product evolves from an innovative to a mature and highly standardized product (Vernon 1966). In the early stage of innovation, production methods are fluid and decisions about the manner of production and the configuration of the innovation must be made quickly and frequently. This need for flexibility requires that production take place in proximity to decisionmakers, a skilled labor force that can adjust to changes in the production process, and agglomeration economies so specialized services can be called in and the activities of competitors monitored.

As the product becomes standardized and production methods routine, ties to decisionmakers, a skilled labor force and agglomeration economies relax, and firms can then begin a nationwide or worldwide scan for new production sites. At this stage, firms can locate branches to take advantage of low-skilled and lower cost labor (Vernon 1966).

The low-wage sites in the United States have been mainly in the South with its unskilled and growing labor force from its own rural population and Mexico and the Caribbean. The textiles and shoe industries moved South in the 1920s and 1930s. Electronics moved South in the 1950s and 1960s and semiconductors moved South in the 1970s and early 1980s. During these periods, plant closures should have been particularly high in these respective industries in the industrialized states and in the case of semiconductors, in California.

Because of intensifying international competition and improved communications and transport, U.S. companies are now shifting the production of these standardized products even further afield to low-wage countries. Another increasingly common way of reducing the costs of inputs is outsourcing, i.e., purchasing components manufactured abroad which were previously produced internally by the firm. In either case, U.S. plants producing with low-skilled labor are expected to be especially vulnerable to closure as they fail to compete with imports or move overseas. Whereas the first two hypotheses predict higher plant closure rates in the northern industrial states, this and the third hypothesis predict higher closure rates in southern states. In terms of the model, this hypothesis explains decreases in TC_o because technological change in production processes and communications have made a new range of alternative production locations possible.

Aside from passing through the product-cycle model, additional characteristics of industries likely to be hard hit by imports and to shift production overseas are those with a high value-to-weight ratio and a high elasticity of demand. A high value-to-weight ratio implies low transport costs as a proportion of product value. A high elasticity of demand implies that small differences in price will make large differences in the size of the market for the product (Vernon 1966).

Empirical work on industrial location and trade patterns offer some support to the product cycle model. In the computer industry (SIC 3572), plants in the vicinity of highly skilled computer workers, such as in the Northeast and northern California, are more involved in the design and development of new computer systems. The production of peripheral computer equipment is dispersing to areas where workers are less skilled and less expensive, such as Tennessee, Nebraska, South Dakota, Utah, North and South Carolina, and Florida (Hekman 1980a). In a second example, new branch plants in the specialized surgical instrument industry were found to be moving into the major centers of current production rather than into the low-wage regions such as the South (Hekman 1980b). Surgical equipment may be an industry which is still early in the product life cycle and still requires a highly skilled labor force, or it may be an industry which never goes through the classic product cycle phases.

A study of the Committee for Economic Development (1984, pp. 92-93) determined the extent to which developing country exports are replacing domestic production in the U.S. Import penetration ratios, measured as the ratio of imports from the developing countries to apparent domestic consumption (domestic production plus imports minus exports) were calculated. The numbers demonstrated that developing country exports constitute only 3 percent of all goods consumed in the U.S., but market penetration has increased rapidly, rising 11.3 percent over the decade of the 1970s. Moreover, this report indicated, as expected, that U.S. imports from developing countries were heavily concentrated in low-wage, labor-intensive industries, with over 50 percent in textiles, apparel, and electrical machinery.

To summarize, because low-wage, low-skilled manufacturing establishments are facing intense foreign competition, we expect profit rates to be relatively low and plant closures likely in import-competing industries and regions dependent on low-skilled, low-wage labor.

Hypothesis 5: Market Demand

The fifth and final hypothesis is that plants faced with declining demand for their product will be most likely to experience falling revenues, shrinking profits, and closure. Markets can be local or national in scope. For example, many manufacturers produce intermediate inputs for local producers, which in turn sell to a national or local market. Other firms in the same industry produce for a national market, and some, of course, produce for both local and national markets. Since we cannot identify the scope of the market for each plant in our study, we hypothesize that growth in both national and local demand affect each plant. We expect to find a higher probability of closure for plants manufacturing products for which there is a shrinking national market and that are located in a declining regional economy. Closures will be least likely for plants located in growing regions and producing products for which there is strong national demand. In terms of the model, plants experiencing a drop in TR_i are more likely to experience negative profits in equation (1) and consequently to close.

Summary of Hypotheses

To summarize, we hypothesize that plant level profits are influenced by the level of local production costs, the extent to which the plant's labor force is unionized, changes in local production costs relative to costs in other regions, the extent to which the plant is competing with foreign imports, and the growth in local and national markets for the plant's product. The probability a plant will be shut down rises as its level of profit falls. The model is the following:

(3) $PC_{i,t-t+1} = f [TC_{r,t}, (U_r/U_o)_t, \Delta(TC_r/TC_o)_{t-t+1}, I_{i,t-t+1}, M_{i,t-t+1})]$.

Where:

PC_i	=The probability plant i will close.
TC	=Total production costs.
U_r/U_o	=Rate of local labor force unionization, relative to the national rate of unionization.
$\Delta(TC_r/TC_o)$	=Change in local production costs relative to the nation.
I	=Import penetration.
M_i	=Change in demand for plant i's output.
r	=Local economy.
t	=Beginning of period.
t+1	=End of period.
i	=Plant.

Empirical Test

The model tests the effects of local production cost levels, unioniza-tion, changes in local production costs relative to costs in other loca-tions, import penetration, and growth in market demand on the prob-ability a plant will close. The model is tested on three industries: metalworking machinery (SIC 354), electronic components (SIC 367), and motor vehicles (SIC 371). The selection of these three industries was discussed in chapter 2. The observations are plants in the Dun and Bradstreet data set in 1975.

The Dependent Variable

The dependent variable takes on value of 1 if the plant was in ex-istence in 1975 and not in existence in 1982. The dependent variable takes on the value of zero if the plant remained open throughout the period. The 1975 to 1982 period was selected to include one full business cycle. The period 1975 to 1979 was an expansionary year for the national economy, and 1979 to 1982 was a recessionary period. The inclusion of one full cycle minimizes any biases in the results that occur due to regional variations in the severity of business cycles (see Howland 1984). For example, the industrialized midwestern states experience more volatile business cycles than the rest of the country because of their dependence on capital goods production. The inclusion of only years that coincided with a recession may make it appear that high wages lead to plant closures, when the true cause of high closures in the Midwest was the business cycle rather than the region's high wages.

The test only includes plant closures that occurred in SMSAs. The reason is as follows. To test the impact of local conditions on plant closures, we need to approximate local labor and energy markets as closely as possible. For example, to test the extent to which high wages increase the probability a plant will close, we need to approximate wages faced by the plant in question. Wage rates in the SMSA where the plant is located are a reasonable measure of wages paid by the plant. Statewide wages are not a satisfactory measure, since wages vary substantially within states, especially within larger states such as California.

Independent Variables

The five hypotheses are tested with the following independent variables.

Production Costs Levels

In order to test hypothesis 1, that plant closures are more probable in high production cost areas, we need to identify production costs that vary across the regions of the United States. Production costs that vary spatially are wages, energy, and taxes. While taxes are not a major expense for the average manufacturer, they are included here because of their role in the debate over the causes of plant closures.

Next to materials, labor costs are the largest expense for the average manufacturing establishment. For example, payroll and supplemental labor costs, including such employee benefits as pension programs, unemployment insurance, health benefits, etc., consume 44 percent of revenue in the metalworking machinery industry, 36 percent of revenue in the electronic components industry, and 18 percent of revenue in the motor vehicle industry (U.S. Department of Commerce 1977b, tables 3a and 3b). No other production expense aside from raw materials is as large.

To demonstrate the spatial variation in wages, table 4.1 reports wages for production workers in 1977 in selected states, by region. While the statistical test of the model uses data on SMSAs, state data are used here to condense information. In 1980, there were 318 SMSAs in the United States, and even though not all have employment in these three industries, wage data presented by SMSA would be unmanageable. Table 4.1 shows that in 1977 the Mid-Atlantic and East North Central states had wages above the national average. The New England, South Atlantic, and East South Central states had wages below the national average.

The SMSA-level wage data used in this study are taken from the *Census of Manufactures* and are average hourly wages at the 3-digit SIC level for production workers. The range of wage values for the industries and SMSAs used in this study are reported in appendix 4.2.

Energy is another input to the production process whose costs vary across the country. Compared to labor, however, energy costs are a small proportion of the average establishment's revenues in the three industries studied here. For example, purchased fuels and purchased electric energy are approximately 1 percent of revenues in the metalworking machinery, electronic components, and motor vehicle industries (U.S. Department of Commerce 1977b, tables 3a and 3b).

Energy costs are highest in the New England and Mid-Atlantic states and lowest in the South and West (see table 4.1). Electricity costs are low in the Pacific region due to this region's abundance of hydroelectric power. The Tennessee Valley Authority is responsible for keeping electricity costs relatively low in the East South Central region. For the range of values for SMSAs used in this study, see appendix 4.2.

Energy costs are measured as the "typical" monthly electricity bills for industrial users in the 300 kilowatt hour (KWH), 120,000 megawatt hour (MWH) category. The 300 KWH refers to the amount of electricity

Table 4.1
Indices of Wages, Electric and Other Fuel Costs,
Taxes, and Union Membership for 26 Selected States by Region

State	1977 wages[a] production workers SIC 354	SIC 367	SIC 371	Electric cost[b] (1975)	Other fuels cost[b] (1975)	Taxes[c] (1977)	% Union labor[d] (1978)
New England							
CT	.97	.91	.68	164.6	128.0	2.42	106.9
MA	.94	.93	NA	175.4	125.7	1.76	119.1
RI	.92	.83	NA	177.7	118.5	1.76	132.3
Mid-Atlantic							
NJ	.97	.93	NA	157.1	127.4	1.82	112.3
NY	1.00	1.10	1.04	124.7	119.6	1.39	191.3
PA	1.02	1.17	NA	125.0	101.1	1.05	166.9
East North Central							
IL	1.00	.91	.80	100.7	98.9	1.12	153.8
IN	.97	.92	.97	88.6	97.8	.66	143.9
MI	1.13	.84	1.12	122.5	110.6	2.45	168.0
OH	1.04	1.06	1.08	87.8	95.5	.73	144.0
WI	1.04	.88	1.03	103.6	103.9	.92	135.7
West North Central							
IA	1.00	.86	.54	98.8	97.8	.49	93.7
MN	.99	1.00	.92	105.3	102.3	1.23	119.1
MO	.94	1.37	NA	101.1	93.9	.64	146.4
South Atlantic							
FL	.81	.85	NA	118.5	103.9	.86	57.1
GA	.83	.80	NA	114.7	90.0	.47	66.4
MD	NA	.79	NA	129.3	122.4	.95	102.5
NC	.75	.89	.58	99.6	105.0	.59	31.8
SC	.67	.80	.42	90.8	101.1	.92	32.7
VA	.99	.78	.75	115.4	108.4	.58	62.0
East South Central							
AL	.81	NA	.64	84.6	87.2	.60	93.7
KY	NA	NA	.97	71.4	93.3	.39	109.3
TN	.82	.87	.62	76.3	86.6	.53	86.4
West South Central							
AR	.90	1.01	.50	81.1	83.8	.47	73.2
TX	.81	1.13	.76	73.5	77.1	.51	53.7
Pacific							
CA	.95	1.03	.85	104.9	98.9	1.80	115.7

SOURCES: *1977 Census of Manufactures*, table 5, volume III, part 1 and 2; U.S. Department of Energy, Energy Information Administration, *Energy Price and Expenditure Report*, 1970-1981, DOE/EIA-0376 (81), June 1984; U.S. Department of Energy, Energy Information Administration, *State Energy Data Report Consumption Estimates 1960-83*, DOE/EIA-0214 (83) section 4, May 1985; William Wheaton (1983) "Interstate Differences in the Level of Business Taxation," *National Tax Journal*, 36, 1, pp. 83-94; and U.S. Department of Labor, Bureau of Labor Statistics, *Handbook of Labor Statistics*, Bulletin 2070, December 1980, table 166.
a. Average Production Wage in state divided by National Average Wage.
b. Index for "other fuels" calculated using data for the states on the price of petroleum, coal, and natural gas in dollars per million BTUs and weighted by the consumption of each fuel by state.
c. Index of business taxes paid by the manufacturing sector as a percent of all manufacturing income.
d. Union membership as a percent of nonagricultural employment by state, 1978.
NA=not available.

used and the 120,000 MWH category refers to the intensity of the power. The important factor here is that energy costs are measured as the bill the same manufacturer would pay in different cities across the country, if its use of energy was exactly the same in each city. We use only electricity costs because of the relative ease of collecting these data for SMSAs and because the major share of energy in our three industries comes from this source. Sixty-eight percent of energy expenditures in metalworking machinery was for electricity. The comparable figures for electronic components and motor vehicles are 78 percent and 57 percent. The remaining shares of energy expenditures are spread among fuel oils, coals, coke and breeze, and natural gas (U.S. Department of Commerce 1977a).

Business taxes are an additional production cost that varies by area of the country. The business tax burden includes property, corporate income, unemployment, payroll, insurance, and stock transfer taxes and general license or business fees. Wheaton (1983) estimates business taxes to be an average of 8 percent of net business income. Taxes as a share of business revenues, the estimate comparable to the wage and energy figures presented above, is not available. In the same paper, Wheaton estimated state tax burdens for 1977 by summing up all business tax payments by each state's manufacturing sector and dividing this value by the state's business income in manufacturing. An index of his estimates is found in table 4.1. In general, the index shows tax burdens to be relatively high in New England and the Mid-Atlantic, and in Michigan, Illinois, and California. The tax burdens in the West North Central, South Atlantic, and East South Central regions are relatively low. The state ratios calculated by Wheaton are used in this study. State tax rates rather than SMSA rates are used because these business taxes are levied at the state level and therefore do not vary within the state. Because wages are the most important input, we should expect closures to be more sensitive to variations in wages than either energy costs or taxes.

Unionization

Rates of union membership also vary spatially, with workers in the industrialized North and West most likely to be represented by unions.

Approximately one-third of nonagricultural employees are represented by unions in these two regions. The lowest rates of union representation are in the West South Central and South Atlantic regions (see table 4.1).

If the level of unionization influences disinvestment in any industry, the three industries studied here would be among them. National data on the proportion of workers covered by unions are not available at the 3-digit level of industrial detail. However, the data at the 2-digit level indicate all three industries may have above-average rates of union participation. Twenty-nine percent of machinery production workers (SIC 35) are unionized. Thirty-six percent of electronics and electrical production machinery workers (SIC 36) are represented by unions and 57 percent of transportation production workers (SIC 37) are union members.

For purposes of this study, unionization is measured as the percent of the nonagricultural labor force belonging to unions. These data are not available at the metropolitan level of disaggregation, and, therefore, the state level data, reported in table 4.1, are used.

Production Cost Changes

Hypothesis 3 predicts greater declines in profit levels, and therefore a greater probability of plant closures, in areas with above-average increases in production costs. Inputs whose costs changed relative to the U.S. average over the period of study include wages and energy costs.

For the three industries studied here, as well as all manufacturing, cross-regional differences in wages have narrowed. The largest relative wage increases have been in the rapidly growing regions of the U.S. Wages in the stagnating or slow-growth regions have dropped relative to the national average wage (see table 4.2). A positive value in table 4.2 signifies an increase in the state's position relative to the U.S. average, while a negative value means a decrease in the state relative to the U.S. average.

Energy price increases over the 1975-81 period have also favored producers in the Mid-Atlantic and some of the East North Central, as well as South Atlantic states. Price increases were greatest in California, Texas, and New Hampshire (table 4.2). The above-average price

Table 4.2

Changes in the Indices for Production Wages 1977-82 and for Electric and Other Fuels Cost 1975-81 for 26 Selected States by Region

State	Change in wages (1977-82)[a] SIC 354	SIC 367	SIC 371	Electric cost change[b] (1975-1981)	Other fuels cost change[c] (1975-1981)
New England					
CT	-.01	-.06	.08	-3.8	5.2
MA	-.04	.01	NA	-4.2	6.0
RI	-.04	-.07	NA	1.4	3.1
Mid-Atlantic					
NJ	-.04	-.01	NA	-2.8	4.2
NY	-.02	-.09	-.30	-5.9	8.2
PA	-.02	.06	NA	-13.1	-9.9
East North Central					
IL	.05	-.03	-.02	2.8	-7.8
IN	-.02	-.04	-.04	4.7	-17.5
MI	.03	-.04	-.03	-7.8	-13.9
OH	.01	-.01	.01	-4.2	-14.5
WI	-.02	NA	.06	-15.6	-3.2
West North Central					
IA	-.05	.04	.07	-8.8	-2.8
MN	-.03	-.11	.04	-10.8	-4.4
MO	-.02	-.03	NA	-10.2	-5.5
South Atlantic					
FL	-.08	0	NA	12.0	-1.0
GA	.01	.06	NA	-27.7	2.9
MD	NA	.09	NA	-29.4	-13.8
NC	.09	.07	.05	-15.9	-15.1
SC	.02	.05	.13	-14.2	-3.7
VA	-.08	.27	NA	-20.8	-0.2
East South Central					
AL	NA	NA	.12	8.5	-7.5
KY	NA	NA	.05	13.7	-15.1
TN	0	.03	.03	15.4	-2.8
West South Central					
AR	.16	.04	.03	-1.3	16.5
TX	.06	-.11	.04	22.3	7.2
Pacific					
CA	.01	.06	-.03	32.3	12.9

SOURCES: *Census of Manufactures,* 1977 and 1982, table 5, vol. III; U.S. Department of Energy, Energy Information Administration, *State Energy Price and Expenditure Report, 1970-1981,* DOE/EIA-0376 (81), June 1984, and U.S. Department of Energy, Energy Information Administration, *State Energy Data Report Consumption Estimates, 1960-1983,* DOE/EIA-0214 (83), section 4, May 1985.

a. These numbers are calculated by subtracting the 1977 Wage Index Value, shown in table 1, from the 1982 index value (not shown).

b. These numbers are calculated by subtracting the 1975 values, shown in table 1, from the 1981 index value (not shown).

c. Index for "other fuels" calculated using data for the states on the price of petroleum, coal, and natural gas in dollars per million BTUs and weighted by the consumption of each fuel by state. NA=not available.

increase in Texas is explained by this state's dependence on natural gas and the decontrol of natural gas prices during this period. While petroleum prices increased at an annual average rate of 18 percent per year, between 1975 and 1981, natural gas increases averaged 22 percent per year. Sixty percent of the West South Central's energy for industrial uses comes from natural gas.

The variables used in this study are the change in the average hourly industry wage between 1977 and 1982. The change in energy costs are the change in a typical electricity bill for an industrial user using 300 KWH in the 120,000 MWH category from 1978 to 1982. Because of the difficulty of locating an index for all business taxes for two periods, and the relatively small proportion of total costs attributable to taxes, the change in tax burden is not included in the model. Again, the range of values for the included SMSAs, is reported in appendix 4.2.

Import Penetration

To measure the extent to which a plant is adversely affected by imports, we selected two proxies. One is the plant's 4-digit line of business. We expect that the greater the import penetration in the plant's 4-digit industry, the greater the loss in revenues and profits, and the greater the probability a plant in that industry will close. Table 4.3 reports the 4-digit industries within metalworking machinery, electronic components, and motor vehicles, and ranks them by the extent to which imports in each industry are replacing domestic production.

Two measures of import competition are reported in table 4.3. The first column, the annual growth rate in imports, measures the rate at which imports are growing. Because this measure does not take into account the growth in domestic demand, it may not accurately represent import penetration. Therefore, table 4.3 also includes a second indicator of import penetration. The second column of table 4.3 measures the growth rate in the ratio of imports to domestic production (value of shipments), weighted by imports' share of value of shipments. This value measures the rate at which imports are replacing domestic production, accounting for the proportion of imports in the industry at the beginning of the period.

Table 4.3
Import Growth Rates by 4-Digit SIC Codes,
Metalworking Machinery, Electronic Components, and Motor Vehicles

SIC code	Annual growth rate in imports (1975-1980)	Growth rate in the ratio of imports to value of shipments, weighted by imports share of value of shipments in 1975 (1975-1980)
Metalworking machinery (SIC 354)		
3541-Metal cutting	12.94	.10
3542-Metal forming	8.71	.07
3546-Power hand tools	11.72	.06
3547-Rolling mill machines	3.64	.04
3544-Special dies	2.10	.01
3549-Metalworking N.E.C.	-3.92	.00
3545-Accessories	7.05	-.72
Electronic components (SIC 367)		
3675-Electronic capacitors	9.35	.07
3674-Semiconductors	15.82	.06
3676-Resistors	13.70	.03
3671-Electronic tubes	8.87	.02
3679-Components N.E.C.	17.69	.02
3677-Coils & transformers	.00	.00
3678-Connectors	2.00	.00
Motor vehicles (SIC 371)		
3711-Bodies	8.07	.12
3714-Parts & accessories	3.43	.03
3713[a]-Truck bodies	.00	.00

SOURCES: Department of Commerce, Bureau of Census, *U.S. Imports, Consumption and General SIC-Based Products by World Areas, 1982*, table 1, May 1983; and *U.S. Imports for Consumption and General Imports, 1975*, table 1, May 1976.

a. Includes motor homes (SIC 3716) along with truck and bus bodies.

Table 4.3 indicates variation in import penetration across 4-digit SIC industries within SICs 354, 367, and 371. For example, in the metalworking machinery industry, imports are cutting into domestic production in all industries except SIC 3549 N.E.C. (not elsewhere classified) and SIC 3545 Accessories. The annual growth rate in imports is greatest in SIC 3541 metal cutting machinery. SICs 3675 and 3711 are experiencing the greatest competition from imports in the remaining two 3-digit industries.

The 4-digit industries enter the model as dummy variables, with SICs 3541, 3671, and 3711 as the reference industries. If import competition is one cause of plant closures, we would expect to see statistically significant and negative values for the coefficients on the dummy variables on industries experiencing less import competition (smaller numbers in column 2 of table 4.3) than that of the reference industries.

A second measure of import competition is the percent of the local labor force with more than a high school education. Even within 4-digit industrial categories, there is still a wide variety of manufacturing activities, some requiring specialized, highly skilled labor and others requiring low-skilled assembly. The low-skilled assembly operations are hypothesized to be most affected by foreign competition and plant closures.

As a proxy for skill level in manufacturing, we use the percent of the SMSA labor force having 12 years of schooling or more. The hypothesis is that plants producing a standardized product with a low-skilled labor force, and consequently vulnerable to imports and closure, will be concentrated in locations with a large labor force with little education. The percent of the labor force with a high school education ranges from 48.2 for New Bedford, Massachusetts to 81.7 percent for Seattle, Washington.

Demand for Output

Three variables are used to measure demand for a plant's output. Two measures of local demand are considered, however this variable is measured differently for each of the three industries. The market for metalworking machinery is primarily other manufacturers. Metalworking machines are generally specialized products designed for a specific manufacturer. Since face-to-face contact is often required, machinery shops are small scale and serve a local clientele (Trainer 1979). Thus, local market growth is expected to play a significant role in the profitability of metalworking machinery plants. This variable in the machine tool equation is measured by the rate of manufacturing employment growth in the plant's SMSA.

The electronic components industry produces intermediate products for local and national markets and final goods primarily for national markets. Although local final demand is not expected to be particularly important for the electronics industry, the percentage increase in population in the plant's SMSA is included here. To capture local intermediate demand, the percentage increase in manufacturing employment in the plant's SMSA is included.

The motor vehicle industry primarily produces final goods for local and national markets. While motor vehicle plants most frequently serve a national market, assembly plants are situated near markets. Therefore, to capture local demand, we include the population growth rate in a plant's SMSA.

For the above variables, manufacturing changes cover the years 1975 to 1982. Population changes cover the years 1970 to 1980. For the SMSAs included in our sample, manufacturing employment growth ranged from –4.85 percent per annum in Youngstown, Pennsylvania to 10.21 percent in San Jose, California. Population growth rates varied between –.9 percent per annum in Jersey City, New Jersey to 5.0 percent per annum in Fort Lauderdale, Florida.

National market demand is captured in the same 4-digit SIC dummy variables reported above under import penetration. This approach was taken to attempt to distinguish between the two hypotheses, given the limited plant specific data. If the dummy variables on industries with strong import penetration are significantly different from zero and positive, then there is support for the import penetration hypothesis. If the dummy variables on industries experiencing the slowest growth are significantly different from zero and of the correct expected sign, then there is support for the hypothesis that national market demand influences closures.

Table 4.4 reports the growth in output (value of shipments) for each of the 4-digit industries. If greater industry growth in output reduces the probability a plant in that industry will close, we would expect positive values on the dummy coefficients for industries growing slower than the reference industry and negative values for the industries growing faster than the reference industry. For example, support for this hypothesis would be indicated by negative values on the dummy coefficients for SICs 3545, 3546, and 3549 and positive values on the dummy coefficients for SICs 3542, 3547, and 3544 in the metalworking machinery industry.

Table 4.4
Growth in Value of Shipments for 4-Digit SIC Categories
Ranked by Growth Rate

SIC code	Growth rate in value shipments, constant dollars, 1970-80 (annual average percent)
Metalworking machinery	
3542 (Metal forming)	.48
3547 (Rolling mill machinery)	.42[b]
3544 (Special dies)	.90
3541 (Metal cutting machinery)	2.88
3545 (Accessories)	4.90
3546 (Power hand tools)	5.50[b]
3549 (N.E.C.)	7.28[b]
Electronic components	
3671 (Electronic tubes)	.05
3677 (Coils and transformers	.43[b]
3676 (Resistors)	1.83[b]
3675 (Electronic capacitors)	4.47[b]
3679 (N.E.C.)	5.93[b]
3678 (Connectors)	10.90[b]
3674 (Semiconductors)	11.47
Motor vehicles	
3711 (Motor vehicles bodies)	−1.95[b]
3714 (Parts and accessories)	.28
3715 (Truck trailers)	4.80
3713 (Trucks and bus bodies)[a]	5.04

SOURCES: *Census of Manufactures, 1977* and *Annual Survey of Manufactures, 1980.*
a. Includes motor homes (SIC 3716).
b. 1972 to 1980 annual average growth rates, 1970 data were not available.

The 4-digit industries exhibiting slow output growth and intense foreign competition are correlated. Therefore, it may not be possible to distinguish between the hypotheses on import penetration and market demand on the basis of this variable. Rather, if import penetration is an important factor in plant closures, we will expect both the 4-digit SIC code and the education variables to be significant.

Control Variables—Size, Age, and Status

As discussed in chapter 3, a plant's size, age, and status are expected to influence the probability of its closure. Larger firms and establishments are expected to close at lower rates than smaller firms and establishments. Younger firms are expected to close at higher rates than well-established firms, and branch plants and subsidiaries are expected to close at higher rates than independents and headquarters. The rationale for including each of these control variables is discussed in depth in chapter 3.

The Impact of Industry Characteristics

As argued above, an industry's rate of technological change, rate of growth in product demand, and market structure should influence its sensitivity to several of the above variables. First, plants experiencing technological change and improvements in labor productivity should be particularly sensitive to unionization in their plant closure patterns. This should result in larger coefficients on the unionization variable in the industries experiencing the greatest rate of technological change.

While all three of the industries studied here (machinery manufacturing (SIC 354), electronic components (SIC 367), and motor vehicles (SIC 371)) are experiencing some employment growth, only electronic components can be characterized as experiencing rapid technological change. Productivity growth in the machine tool industry has been sluggish, with 1 to 2 percent annual increases in output per employee hour over the period 1975 to 1982 (Belitsky 1983). The reason for little technological adoption of new production processes in this industry is that most firms are small, single-plant operations that cannot afford expensive new capital equipment. For example, in SIC 3544, metal cutting machinery, there are 7100 establishments with an average of 15 employees in each. Numerically controlled machine tools, the major technological advance in machine tool production, would replace labor, reduce scrap, and increase precision, but they are too costly to be affordable for most machine tool firms. Consequently, only 3 percent of metal cutting machine tools are numerically controlled (Belitsky 1983).

While the electronic components (SIC 367) group includes diverse industries, overall technological change has been rapid, leading to large increases in worker productivity. Over the period 1967 to 1980, output per employee hour increased by 8.8 percent annually in the electronics industry (Critchlow 1983b). Improvements in assembly line procedures and automation of many tasks are changing the skill requirements and increasing worker productivity. Assembly jobs, the largest occupational category, are being lost to machine monitoring, feeding, loading, unloading, and maintenance. Demand for solderers, technicians, and drafters has also declined, while demand for scientists and engineers has increased.

Technological change in the auto industry has been slow, relative to that of electronics. From 1975 to 1982, output per employee hour for production employees increased by 1.8 percent per year (U.S. Department of Labor 1983, p. 219). While this industry has mechanized much of its assembly line operations, the most common technological changes over the period of study here have been in computerized auto styling and design and computerizing inventory management, such as using computers to keep track of parts and production materials, and forecasting parts shortages.

Earlier, we hypothesized that firms adopting new technologies may find it is easier to close existing plants and adopt new production processes in a new plant, rather than attempt to accommodate new technologies in an existing unionized plant. If true, we would expect closures in electronic components plants to be more sensitive to unionization than closures in metalworking machinery and motor vehicles. Support for this hypothesis would show up in larger coefficients on the unionization variable in the electronic components equation than the motor vehicle and machine tool equations.

Second, industry rates of growth are also expected to influence regional plant closure patterns. Reductions in capacity are expected to be concentrated in plants located in high-cost regions. In fast-growth industries, high-cost plants are also expected to close at higher rates than low-cost plants, but because the capacity at high-cost plants may be needed to meet excess demand, the closure rates are not expected to be as high as for declining industries. These differences should show up as larger coefficients on the production cost variables in a declining (motor vehicles) than a growing (electronic components) industry.

Third, cross-industry differences in market structure should also influence the size of the coefficients across the three industries studied here. All three industries studied here face competition for markets and therefore are expected to be responsive to the variables outlined above. However, in the case of motor vehicles, which was an oligopoly until the late 1970s, plant closures are expected to be particularly sensitive to high production costs and unionization.

Markusen (1985) argues that when oligopolies emerge early in the evolution of an industrial sector, pricing practices, restrictions in output, and product differentiation result in the spatial concentration of production. This spatial concentration is not the result of a least cost calculus, as Markusen demonstrates for the case of steel, and exists because higher costs are easily passed on to consumers. Once the oligopolist's market is eroded, either through the development of substitutes, or, as in the case of autos, through competition from abroad, "oligopolists have the incentive, resources, and know-how to rationalize and restructure very rapidly with devastating consequences for traditional centers" (p. 12). If Markusen's argument holds for the auto industry, we would expect to see very high closure rates in the high-cost, unionized plants, which are located in the traditional centers of automobile production. This would show up as large, positive, and statistically significant coefficients on the variables for labor costs, energy costs, and taxes.

Summary of the Equation to be Estimated

Five hypotheses are to be tested with 16 independent variables. An establishment is more likely to close (1) the higher the level of labor, tax, and utility costs, (2) the more unionized the local labor force, (3) the greater the increase in local labor and energy costs relative to the national average cost increase, (4) the lower the percent of the labor force with a high school degree and the greater the competition from imports, and (5) the greater the decline in product demand in local and national markets. Several additional variables that influence the probability a plant will close, including the plant's status as an independent, headquarters, branch, or subsidiary, the plant's age, and the plant's size, are included in the model to hold these factors constant. The final equation to be estimated is:

$$\begin{aligned}
(4) \quad PC_i = {} & \alpha_0 + \alpha_1 Wage_s + \alpha_2 Energy_s + \alpha_3 Tax_s \\
& + \alpha_4 Union_s + \alpha_5 ChWage_s + \alpha_6 ChEnergy_s \\
& + \alpha_7 HS_s + \alpha_8 I_i + \alpha_9 Manu_s + \alpha_{10} PopCh_s \\
& + \alpha_{11} Hdquarters_i + \alpha_{12} Subsidiary_i \\
& + \alpha_{13} Branch_i + \alpha_{14} Size_i + \alpha_{15} Age_i \\
& + \alpha_{16} New_i
\end{aligned}$$

where

PC = A binary variable, where PC = 1 if the plant closed between 1975 and 1982 and 0 if the plant remained in business.

Wage = Average hourly wage rate for production workers in the 3-digit SIC code industry between 1977 and 1982.

Energy = A "typical" monthly utility bill between 1975 and 1981.

Tax = Percent of business income spent on business taxes in 1977.

Union = Percent of the nonagricultural labor force in unions, 1978.

ChWage = Change in wages, 1977 to 1982.

ChEnergy = Change in energy costs, 1975 to 1981.

HS = Percent of the labor force with a high school degree, 1980.

I = Dummy variable representing the industry's 4-digit SIC code.

Manu = Percentage change in manufacturing employment, 1975 to 1982.

PopCh = Percentage change in population, 1970 to 1980.

Hdquarters = Dummy variable with 1 = headquarters and 0 if not.

Subsidiary = Dummy variable with 1 = subsidiary and 0 if not.

Branch = Dummy variable with 1 = branch and 0 if not.

Size = Size of establishment in number of employees.

Age = Age of establishment (only available for single-plant firms).

New = Dummy variable if firm was established after 1969 (only available for single-plant firms).

s = Standard Metropolitan Statistical Area.

i = Plant.

Results

The results are reported in tables 4.5 through 4.7. The equations for each industry were estimated for all establishments and for branches and independents separately. Branches and independents were estimated separately for several reasons. First, it allows for the possibility that the slopes of the coefficients vary by type of establishment. For example, the Rhode Island Strategic Development Commission (1983) survey found closures of branches to be more sensitive to production costs than closures of independents. Second, separate equations allow us to check for biases caused by faulty data. As shown in chapter 2, there is evidence the data on the dependent variable is more accurate for branches than for independents. Third, a separate model for independents (single-plant firms) also permits the effects of a firm's age to be held constant. As noted above, the D&B data do not include the age of branch plants.

To capture more homogeneous industrial categories, the establishments used in the following models are only those that identified the given industry as its first line of business. There were 46 SMSAs with metalworking machinery establishments for which a complete set of data could be compiled. There were a total of 6,303 establishments in these 46 SMSAs in 1975, 1,573 of which closed over the 1975 to 1982 period. The results for the electronic components industry (table 4.6) are based on 1,083 establishments in 36 SMSAs. Of these 1,083 establishments, 397 closed over the 1975 to 1982 period. There were 23 SMSAs with motor vehicle establishments and a complete set of data. There were 1,358 establishments in these 23 SMSAs, 457 of which closed over the period under study. See appendix 4.3 for a list of the SMSAs included in the analysis and the number of plants in each SMSA.

The findings demonstrate little support for the hypotheses that local economic variables are responsible for plant closures. For example, in the machine tool industry equation, the only coefficient statistically significant at the 5 percent level of confidence is the manufacturing growth rate (Manu), and the sign is positive rather than the expected negative. A positive sign indicates that the faster the growth in local manufacturing employment, the more likely a metalworking establishment will close, a result inconsistent with the hypotheses set forth earlier.

In logit analysis, the coefficients cannot be directly interpreted. Therefore, the elasticities (the increase in the probability of a closure given a 100 percent increase in the independent variable) are calculated for the significant variables. These elasticities are relevant only when all other variables take on their mean value. For example, the value in the brackets [   ] under the coefficient on the local manufacturing growth rate (Manu) can be interpreted to mean that a 100 percent increase in the manufacturing growth rate leads to a 1 percent increase in the probability a plant will close, when all other variables are at their mean value.

The one variable that is significant at the 10 percent level and of the correct sign is wages (Wage) in the equation for independent electronic components firms (see table 4.6). Independent electronic component firms are more likely to close in areas where wage levels are high. The value in brackets is interpreted to mean that a 100 percent increase in the wage level results in a 12 percent increase in the probability an independent will close, when all other independent variables are at their mean value.

In the cases where other variables are statistically significant, the signs are the opposite of those expected. For example, table 4.7 reports a positive sign on the coefficient for the percent of the population with a high school degree, indicating a greater probability of a motor vehicle plant closing where the labor force is better educated. In addition, the same table shows a negative but significant sign on the coefficient for taxes in the equation for independents. This indicates that the probability of a firm closing is lower in high-tax SMSAs.

Further, there are no statistically significant differences in the probability of a plant's closing across 4-digit industries, even though there are wide variations in the extent to which imports are cutting into the domestic production and wide variations in their rates of output growth (see tables 4.3 and 4.4). For example, if serious import penetration

Table 4.5
Effects of Economic Variables on Plant Closures
Metalworking Machinery (SIC 354)
Logit Model

	All	Branches	Independents
Constant	.88	−2.99	−.23
	(1.04)	(.72)	(.06)
Wage	−.08	.15	−.09*
	(2.56)	(.50)	(3.42)
			[−.02]
Energy	.00	.00	.00
	(2.46)	(.31)	(.43)
Tax	.59	2.54	.91
	(.58)	(.80)	(1.16)
Union	.01	−.00	.01
	(.33)	(.00)	(.85)
ChWage	−.08**	−.21	−.07
	(2.67)	(.83)	(2.27)
	[−.02]		
ChEnergy	−.00	.00	−.00
	(.54)	(.02)	(.03)
HS	.00	.02	−.00
	(.11)	(.32)	(.13)
SIC 3542	.11	.25	.40
(Metal forming machine tools)	(.12)	(.16)	(.76)
SIC 3544 (Special dies)	−.29	−.14	−.07
	(.83)	(.07)	(.02)
SIC 3545	−.22	−.24	.02
(Machine tool accessories)	(.47)	(.18)	(.00)
SIC 3546	−.51	−.57	−.14
(Power driven hand tools)	(1.69)	(.68)	(.07)
Manu	.05*	.03	.05*
	(5.88)	(.22)	(5.48)
	[.01]		[.01]

Table 4.5 (continued)

	All	Branches	Independents
Hdquarters	.05 (.14)	—	—
Subsidiary	.76* (20.94) [.17]	—	—
Branch	1.35* (141.46) [.32]	—	—
Size	−.002* (10.29) [−.003]	.00 (1.09)	−.02* (75.35) [−.004]
Age	—	—	.001** (3.23) [.001]
New	—	—	.27* (13.32) [.01]
No. of observations	6303	404	5554
Fraction of pairs where predicted=actual response	.56	.55	.62

* Statistically significant at the 5 percent level of confidence or greater.

** Statistically significant at the 10 percent level of confidence.

( ) Chi-square statistics.

[ ] The increase in the probability of a plant closing occurring given a 100 percent increase in the corresponding independent variable.

Table 4.6
Effects of Economic Variables on Plant Closures
Electronic Components (SIC 367)
Logit Model

	All	Branches	Independent
Constant	-2.94**	-5.99	-3.99**
	(2.88)	(0.8)	(3.58)
Wage	.35	.39	.51**
	(2.10)	(.52)	(3.08)
			[.12]
Energy	.00	-.00	.00
	(.08)	(.91)	(.67)
Tax	1.70	9.04	2.91
	(.39)	(1.58)	(.82)
Union	-.01	-.01	-.03
	(.60)	(.03)	(1.79)
ChWage	-.32	-.04	-.50
	(1.84)	(.01)	(2.68)
Ch Energy	.00	.00	-.00
	(.09)	(1.34)	(.00)
HS	.00	-.05	.01
	(.04)	(1.46)	(.63)
SIC 3672	.26	7.20	-.06
(Cathode ray & picuture tubes)	(.18)	(.12)	(.01)
SIC 3673	.94	7.41	.15
(Electronic tubes)	(2.62)	(.12)	(.05)
SIC 3674	.81	7.26	.30
(Semiconductors)	(2.45)	(.12)	(.30)
SIC 3675	.79	9.42	-.13
(Electronic capacitors)	(1.66)	(.20)	(.03)
SIC 3676	.36	7.42	-.39
(Electronic resistors)	(.30)	(.12)	(.27)
SIC 3677	-.48	6.86	-1.06
(Electronic coils)	(.64)	(.10)	(2.62)

Table 4.6 (continued)

	All	**Branches**	**Independents**
Manu	−.01	−.01	−.02
	(.05)	(.03)	(.15)
PopCh	.08	.09	.10
	(.70)	(.21)	(.77)
Hdquarters	−.23	—	—
	(.84)		
Subsidiary	.59*	—	—
	(5.64)		
	[.11]		
Branch	.79*	—	—
	(20.01)		
	[.15]		
Size	−.00	−.00	−.002
	(1.93)	(.19)	(2.19)
Age	—	—	.01
			(1.06)
New	—	—	.42*
			(5.81)
			[.11]
No. of observations	1083	186	778
Fraction of pairs where predicted = actual response	.63	.63	.63

* Statistically significant at the 5 percent level of confidence or greater.

** Statistically significant at the 10 percent level of confidence.

( ) Chi-square statistics.

[ ] The increase in the probability of a plant closing occurring given a 100 percent increase in the corresponding independent variable.

Table 4.7
Effects of Economic Variables on Plant Closures
Motor Vehicles (SIC 371)
Logit Model

	All	Branches	Independents
Constant	−3.82*	−6.94*	−4.37*
	(7.53)	(4.47)	(6.09)
Wage	.03	−.15	.03
	(.33)	(1.58)	(.25)
Energy	.000	.001	.000
	(.47)	(1.28)	(1.06)
Tax	−2.75	−4.56	−5.91*
	(1.66)	(1.28)	(4.16)
			[.20]
Union	.004	−.01	.007
	(.08)	(.06)	(.15)
ChWage	−.004	.47	.08
	(.00)	(2.56)	(.23)
ChEnergy	.000	−.000	−.000
	(.00)	(.02)	(1.85)
HS	.03*	.06	.03**
	(4.53)	(2.47)	(2.74)
	[.01]		[.01]
SIC 3713	−.09	.78	−.33
(Truck and bus bodies)	(.11)	(1.75)	(.83)
SIC 3714	.17	.09	.08
(Motor vehicle parts	(.52)	(.04)	(.07)
and accessories)			
PopCh	−.06	−.29	−.08
	(.43)	(1.56)	(.44)
Hdquarters	−.27	—	—
	(1.75)		
Subsidiary	.40**	—	—
	(3.07)		
	[.08]		

Table 4.7 (continued)

	All	Branches	Independents
Branch	1.09* (48.80) [.25]	—	—
Size	−.0003* (5.04) [−.0001]	−.000 (2.60)	−.001 (.32)
Age			.001 (.77)
New	—	—	.73* (20.97) [.14]
No. of observations	1358	264	900
Fraction of pairs where predicted = actual responses	.60	.60	.60

* Statistically significant at the 5 percent level of confidence or greater.

** Statistically significant at the 10 percent level of confidence.

( ) Chi-square statistics.

[ ] The increase in the probability of a plant closing occurring given a 100 percent increase in the corresponding independent variable.

resulted in high industry closing rates we would expect industry 3711 (motor vehicles and car bodies) to experience high rates of closure. This would show up in table 4.7 as negative and significant coefficients for SIC 3713 (truck and bus bodies and motor homes) and SIC 3714 (motor vehicle parts), since these latter two industries are less affected by import competition than SIC 3711.

The absence of evidence of an association between industry closure rates and industry growth rates is similar to findings from plant closing studies in Great Britain. Healey (1982), in a study of the textile and clothing industries, found no relationship between closure rates and industry growth rates, and in studies of Ireland (O'Farrell 1976), the East Midlands (Gudgin 1978), and Scotland (Henderson 1979), researchers found no significant correlation between the same variables.

There are at least three possible explanations for this result. One, low rates of start-ups and expansions may pick up the adjustments when an industry declines, much the same pattern as that shown for regional growth in chapter 3. A second, though not mutually exclusive, explanation is that there are differences across industries in the adjustment to decline and no relationship between the variables the rate of plant closures and industry growth exists. For example, healthy firms in growing industries may exhibit high rates of closure because they have sufficient profits to invest in research and development and to adopt new technologies. The ability to adopt new technologies may lead to plant closures as older, redundant plants are closed. These high closure rates in fast-growth industries may offset a high rate in declining industries.

In all three industries, a plant's status has the greatest impact on the probability a plant will close. Headquarters close at about the same rate as independents. In the metalworking machinery industry, the probability a branch will close is 32 percent greater than the probability a single-plant firm will close. The probability of a subsidiary closing (including headquarters and branches) is 17 percent higher than for an independent. Both of these variables are significant at high levels of confidence in all three equations. As chapter 2 pointed out, the data overestimate closures in independents and give a more accurate reading of branch plant than independent closures. Even given the overestimate of independent closures, the coefficient on branch plants is positive, highly significant, and large.

Size also has an impact on the probability a plant will close, with larger establishments in both the metalworking machinery and motor vehicle industries less likely to close. Interpreting the result from the metalworking machinery industry, every 100 percent increase in size reduces the probability a plant will close by 0.3 percent. The findings on size for this industry can be explained by the tendency for large independents to be less likely to close. This variable is insignificant for branches (table 4.5).

Tables 4.5 through 4.7 also indicate new firms are more likely to close than older firms. New firms are defined as those started after 1969. Only in the metalworking machinery industry is there evidence older firms are more likely to close than younger firms, where every 100 percent increase in age increases the probability a machine tool firm will close by 0.1 percent. The significance and positive values for both (Age) and (New) in the metalworking machinery equation for independents indicates the impact of age on the probability a firm will close is a "U"-shaped function with high probabilities of closure for very young and very old metalworking manufacturing firms, and lower probabilities for middle-aged metalworking machinery firms.

In both the electronic components and motor vehicle equations, very young plants are most likely to close. However, there is no evidence that age influences the likelihood of closure for firms older than five years. These results can be explained by the tendency for metalworking machinery firms to be small single-plant operations that close when the owner retires. The other two industries are comprised primarily of firms with a corporate form of management.

The dependent variable in the previous models included both closings and out-of-state relocations. We estimate a similar model, including only relocations. The number of relocations was too small in both the electronic components and motor vehicle industries to yield meaningful results. However, the model could be estimated for the metalworking machinery industry. These results are reported in table 4.8, and indicate headquarters and subsidiaries and plants located where labor unions are well organized are the most likely to relocate to another state. This finding supports the hypothesis that plants leave regions to avoid dealing with a unionized labor force.

Table 4.8
Effects of Economic Variables on Relocations
Metalworking Machinery (SIC 354), Logit Model

Constant	.47	SIC 3545	
	(.01)	(Machine tool accessories)	−1.26
Wage	−.19		(1.28)
	(.36)	SIC 3546	
Energy	−.00	(Power driven hand tools)	−.61
	(.73)		(.18)
Tax	−3.38	Manu	.10
	(.40)		(.57)
Union	.10*	Hdquarters	1.60*
	(3.72)		(12.20)
	[.03]		[.39]
ChWage	−.03	Branch	−.55
	(.01)		(.28)
ChEnergy	−.00	Subsidiary	1.30*
	(.00)		(4.22)
			[.31]
HS	−.11	Size	−.00
	(2.41)		(.00)
SIC 3542			
Metal forming machine tools)	−.80	No. of observations	6303
	(.47)	Fraction of pairs where	
SIC 3544		predicted=actual response	.55
(Special dies)	−1.35		
	(1.58)		

* Statistically significant at the 5 percent level of confidence or greater.
( ) Chi-square statistics.
[ ] The increase in the probability of an out of state relocation given a 100 percent increase in the corresponding independent variables.

The finding that headquarters are more likely to relocate than branches (table 4.8) should be interpreted judiciously, since D&B is much more likely to record a relocating headquarters correctly than a relocating branch. As discussed in chapter 2, many relocating branches are probably recorded as a closing and a subsequent start-up in a new location. Table 4.8 offers no support, however, for the frequently argued position that high wages, taxes, and energy bills are driving plants out of the northern states.

Conclusions

The evidence from the metalworking machinery, electronic components, and motor vehicles industries does not support the hypotheses that regional differences in production cost levels, level of unionization, production cost changes, import penetration, and local market growth lead to regional differences in rates of plant closures. This result helps explain the absence of regional variations in plant closure rates and rates of job loss reported in chapter 3. There are several possible explanations.

One explanation is that the independent variables affect a plant's profitability with a lag. We measured the level and change in variables over the period 1975 to 1982, the same years as covered by the plant closing data. The plant closing decision may reflect economic conditions from an earlier time period.

A second explanation is that plant closing decisions are not affected by the characteristics of the regions where the plants are located and that the decision to close a plant, once the plant's status, age, and size are held constant is an idiosyncratic event or affected by factors that occur randomly across regions. These latter two conclusions are consistent with the case studies conducted by Massey and Meegan (1982) in Great Britain.

A third explanation for the finding that the characteristics of a plant's location do not have an impact on the plant closing decision is that, unlike the decision to open a plant, the decision to close a plant may be influenced by political and longer run economic variables. For example, when there are other options for cutting back on capacity, a manager may decide against closing a heavily unionized plant or a plant in a depressed community because of fear of adverse attention from the media or consumers, or because the owner feels a responsibility to the community. Schmenner found occasions where closing decisions were made on grounds other than short-run profitability:

> With both costs and with social responsibility in mind, some companies have modified their decisions regarding plant closings from what might be termed a rigid financial accounting approach. The most common modification—one that is seen as the most equitable one amid a distasteful situation—is to close the least senior plant (1982, p. 236).

This strategy guided Texas Instrument's only plant closing during the 1973-75 recession in Ft. Walton Beach, Florida.

A fourth possibility is that the model misses an important intervening variable, i.e., competition from other producers. Evidence from chapter 3 demonstrates that plants located in regions with shrinking markets and high production costs are competing in a less intense environment, reducing the likelihood a plant will close. As profit rates in a region (and/or industry) fall, new investment is deflected to lower cost regions or to higher profit industries (Engle 1974; Carlton 1979). Entrepreneurs shift their attention away from the less profitable market, leaving a less competitive environment to existing producers.

In their case studies of job loss, Massey and Meegan (1982) found two instances where competition was an important consideration in the plant closing decision. Iron casting foundries located in regions with little competition operated relatively inefficient plants. Plants located in regions with more intense competition were more efficient, but also more likely to close. They report a similar finding for the newsprint industry. Newsprint companies, faced with excess capacity, chose to adjust by closing plants located in competitive markets and to continue operating plants where the company held a monopoly position. Market competition was not included in the model presented in this chapter.

The characteristics of regional economies have been a good predictor of where new firms locate. Plants locate where wages and energy costs are relatively low, an appropriately skilled labor force is available, and agglomeration economies exist. Regional characteristics do not, however, appear to influence the geography of plant closures. The only factors that appear to influence the closure decision are attributes of a plant, such as its status, age, and size. The findings of chapter 4 reinforce the conclusion of chapter 3, that plant closing rates are uncorrelated with economic growth. Spatial variations in economic growth appear to be explained by high plant start-up and expansion rates in the fast-growth regions, rather than low rates of plant closures.

Because plant closures occur at roughly even rates in all regions of the country, worker displacement may be a serious issue in the fast-growth as well as the slow-growth economies. In chapter 5, we explore the extent to which regions with strong employment growth absorb the workers permanently laid off. Do displaced workers face difficult labor market adjustment in all regions of the country, or is displacement primarily a problem concentrated in labor markets where there is insufficient job creation?

Appendix 4.1
Sources of Data

Average Hourly Wages by Industry, 1977 and 1982: U.S. Department of Commerce, Bureau of the Census, *Census of Manufactures, 1977*, table 6 and *Census of Manufactures, 1982*, table 6.

Union Membership, 1976: U.S. Department of Commerce, Bureau of the Census, *Statistical Abstract*, 1979, table no. 705, p. 427.

Electricity Rates, 1978 and 1982: United States Department of Energy, Energy Information Administration, *Typical Electric Bills, January 1, 1978*, table 11, August 1978, pp. 166-185; and *Typical Electric Bills, January 1, 1982*, table 19, October 1982, pp. 284-309.

Years of High School and College Completed, 1980: U.S. Department of Commerce, Bureau of the Census, *Population Census*, 1980, table 119, Educational Characteristics for Areas and Places.

Business Taxes on Manufacturing, 1977: Wheaton, William (1983) "Interstate Differences in the Level of Business Taxation," *National Tax Journal*, 36,1, March, p. 89.

Population Changes, 1970-80: U.S. Department of Commerce, Bureau of the Census, *State and Metropolitan Area Data Book, 1982*, table A.

Change in Manufacturing Employment, 1975 to 1982: U.S. Department of Labor, Bureau of Labor Statistics, *Employment and Earnings*, 22,11, May 1976, table 1, pp. 126-134 and *Employment and Earnings*, 31,5, May 1984, table 1, pp. 125-140.

Appendix 4.2
Range of Values of Independent Variables
for Equations Reported in Tables 4.5 through 4.7

Variable	Minimum		Maximum	
Wage SIC 354	5.74	(Fort Lauderdale)	11.62	(Buffalo)
Wage SIC 367	4.54	(New Haven)	8.12	(St. Louis)
Wage SIC 371	5.63	(Memphis)	12.14	(San Francisco-Oakland)
Energy	1218.00	(Seattle)	7519.00	(Boston)
ChWage SIC 354	−1.04	(Buffalo)	5.50	(Denver)
ChWage SIC 367	.42	(West Palm Beach)	4.25	(St. Louis)
ChWage SIC 371	.95	(Boston)	7.06	(San Francisco-Oakland)
ChEnergy	−4548.00	(Fresno)	4266.00	(Hartford)

Appendix 4.3

Number of Establishments Included in Logit Model and Their Regional Distribution

	Number of establishments		
	Metalworking machinery SIC 354	Electronic components SIC 367	Motor vehicles SIC 371
New England			
Boston		95	29
Bridgeport	96	7	
Brockton		4	
Hartford	113		
Lawrence-Haverhill, MA-NH		9	
Lowell, MA-NH		6	
New Britain	61		
New Haven-West Haven, CT		8	
Providence-Warwick-Pawtucket	53	9	
Middle Atlantic			
Allentown	12		
Buffalo	53	5	
Erie	35		
Harrisburg	22		
New York, NY-NJ	407	89	88
Newark	243	37	22
Paterson—Clifton-Passaic, NJ	116	25	
Philadelphia	241	45	69
Pittsburgh	94	7	
Syracuse	18		
York, PA	26		
East North Central			
Akron	106		
Canton	28		
Chicago	947	83	
Cincinnati, OH-KY-IN	101		
Cleveland	373		81
Columbus	21		17
Detroit	1270	16	243
Evansville	33		
Indianapolis	144	4	
Kalamazoo-Portage, MI	21		
Lima, OH			10
Milwaukee	183	12	
Muncie	35		

Appendix 4.3 (continued)

	Number of establishments		
	Metalworking machinery SIC 354	Electronic components SIC 367	Motor vehicles SIC 371
East North Central (cont.)			
Racine	43		
South Bend	41		
Toledo	112		19
Youngstown-Warren	48		
West North Central			
Minneapolis-St.Paul	113	16	35
St. Louis	122	9	41
South Atlantic			
Baltimore		9	
Charlotte-Gastonia, NC			9
Fort Lauderdale-Hollywood	42	9	
Greensboro—Winston-Salem—High Point	13		
Miami	36	6	
Raleigh-Durham, NC		1	
Tampa-St. Petersburg	32	7	
Washington, DC		22	
West Palm Beach-Boca Raton		6	
East South Central			
Louisville, KY-IN			9
Memphis			20
Nashville	35		16
West South Central			
Dallas-Fort Worth	57	37	55
Houston	41		27
Oklahoma City			28
San Antonio			15
Mountain			
Denver	40	19	
Phoenix		17	
Pacific			
Anaheim-Santa Ana-Garden Grove		70	82
Los Angeles-Long Beach	505	177	341
Portland, OR	39		40
San Diego	25	39	
San Francisco-Oakland	78	27	62
San Jose	29	135	
Santa Barbara-Santa Maria-Lompoc		10	
Seattle-Everett		10	
SUM	6303	1087	1358

REFERENCES

Belitsky, Harvey A. (1983) "Technology and Labor in Metalworking Machinery," in *A BLS Reader on Productivity,* U.S. Department of Labor, Bureau of Labor Statistics. Washington, D.C.: Government Printing Office, June, pp. 177-189

Bluestone, Barry and Bennett Harrison (1982) *The Deindustrialization of America.* New York: Basic Books.

Browne, Lynne (1984) "How Different are Regional Wages? A Second Look," *New England Economic Review* (March/April), pp. 40-47.

Carlton, Dennis (1979) "Why New Firms Locate Where They Do?" in William Wheaton (ed.), *Interregional Movements and Regional Growth.* Washington, D.C.: Urban Institute, pp. 13-50.

Committee for Economic Development (1984) *Strategy for U.S. Industry Competitiveness: A Statement Prepared by the Research and Policy Committee for Economic Development.* Washington, D.C.: Committee for Economic Development, April.

Critchlow, Robert V. (1983a) "Technology and Labor in Motor Vehicles and Equipment," in *A BLS Reader on Productivity.* U.S. Department of Labor, Bureau of Labor Statistics. Washington, D.C.: Government Printing Office, June, pp. 190-196.

Critchlow, Robert V. (1983b) "Technology and Labor in Electrical and Electronic Equipment," in *A BLS Reader on Productivity.* U.S. Department of Labor, Bureau of Labor Statistics. Washington, D.C.: Government Printing Office, June, pp. 150-159.

Engle, Robert (1974) "A Disequilibrium Model of Regional Investment," *Journal of Regional Science* 14, 3, pp. 367-376.

Fecteau, George and John Mera (1981) "It's a Walkover for Shoe Imports," in Robert Baldwin and J. David Richardson (eds.) *International Trade and Finance.* Boston: Little, Brown, pp. 87-93.

Finley, Murray (1981) "Foreign Trade and United States Employment," in Robert Baldwin and J. David Richardson (eds.) *International Trade and Finance.* Boston: Little, Brown, pp. 77-87.

Gainer, William J. (1986) "U.S. Business Closures and Permanent Layoffs During 1983 and 1984," Paper presented at the Office of Technology Assessment Workshop on Plant Closings, April 30–May 1.

Gudgin, G. (1978) *Industrial Location Processes and Regional Employment Growth.* Farnborough, England: Saxon House.

Harris, Candee (1986) "High Technology Entrepreneurship in Metropolitan Industries" in Edward Bergman (ed.), *Local Economies in Transition:*

Policy Realities and Development Potentials. Durham, North Carolina: Duke University Press.

Healey, M.J. (1982) "Plant Closures in Multiplant Enterprises—The Case of a Declining Industrial Sector," *Regional Studies* 16, 1, pp. 37-51.

Hekman, John (1980a) "The Future of High Technology Industry in New England: A Case Study of Computers," *New England Economic Review* (January/February), pp. 5-17.

Hekman, John (1980b) "Can New England Hold onto its High Technology Industry," *New England Economic Review* (March/April), pp. 35-44.

Hekman, John (1980c) "The Product Cycle and New England Textiles," *Quarterly Journal of Economics* 94, 4, pp. 697-717.

Henderson, R.A. (1979) "An Analysis of Closures Amongst Scottish Manufacturing Plants," ESU Discussion Paper 3. Edinburgh, Scotland: Scottish Economic Planning Department.

Howland, Marie (1984) "Regional Variations in Cyclical Employment," *Environment and Planning A* 16, pp. 863-877.

Howland, Marie (1985) "Property Taxes and the Birth and Intraregional Location of New Firms," *Journal of Planning Education and Research* 4, 3, pp. 148-156.

Hulten, Charles and Robert Schwab (1983) *Regional Productivity Growth in U.S. Manufacturing.* Washington, D.C.: Urban Institute.

Hymer, Steve (1972) "The Multinational Corporation and the Law of Uneven Development," in J.N. Bhagwati (ed.), *Economics and World Order from the 1970's to the 1990's.* New York: Macmillan.

Kuznets, Simon (1953) *Economic Change.* New York: W.W. Norton.

Leary, Thomas J. (1985) "Deindustrialization, Plant Closing Laws, and the States," *State Government* 58, 3 (Fall), pp. 113-118.

Malecki, Edward J. (1981) "Recent Trends in the Location of Industrial Research and Development: Regional Development Implications for the United States," in J. Rees, G.J.D. Hewings, and H.A. Stafford (eds.), *Industrial Location and Regional Systems.* New York: Bergin Press.

Malecki, Edward J. (1986) "Industrial Locations and Corporate Organization in High Technology Industry," *Economic Geography* 61, 4 (October), pp. 345-369.

Mansfield, Edwin (1985) *Microeconomics: Theory and Applications,* shorter fifth edition. New York: W.W. Norton.

Markusen, Ann (1985) *Profit Cycles, Oligopoly, and Regional Development.* Cambridge, Massachusetts: MIT Press.

Massey, Doreen and Richard Meegan (1982) *The Anatomy of Job Loss.* New York: Methuen Press.

O'Farrell, P.N. (1976) "An Analysis of Industrial Closures: Irish Experience 1960-1973," *Regional Studies* 10, 4, pp. 433-448.

Phillips, Kevin (1984) "Look Out for Number One," *Washington Post* (December 16) Fl.

Reamer, A. (1987) *Plant Closures in Rhode Island,* unpublished Ph.D. thesis, Department of Urban Studies and Planning, Massachusetts Institute of Technology.

Rees, John (1979) "Regional Industrial Shifts in the U.S. and the Internal Generation of Manufacturing in Growth Centers of the Southwest," in William C. Wheaton (ed.), *Interregional Movements and Regional Growth.* Washington, D.C.: Urban Institute, pp. 51-73.

Rhode Island Strategic Development Commission (1983) *The Greenhouse Compact,* executive summary. Providence, Rhode Island: Rhode Island Strategic Development Commission.

Schmenner, Roger (1982) *Making Business Location Decisions.* Englewood Cliffs, New Jersey: Prentice Hall.

Smith, David M. (1981) *Industrial Location: An Economic Geographic Analysis,* second edition. New York: Wiley.

Stafford, Howard (1980) *Principles of Industrial Location.* Atlanta, Georgia: Conway Publications.

Trainer, Glynnis (1979) *The Metal Working Machinery Industry in New England,* unpublished Masters thesis, Department of Urban Studies and Planning, Massachusetts Institute of Technology.

U.S. Department of Commerce, Bureau of the Census (1976) *U.S. Imports, Consumption and General SIC-based Products by World Areas, Annual 1975.* Washington, D.C.: Government Printing Office.

U.S. Department of Commerce, Bureau of the Census (1983) *U.S. Imports, Consumption and General SIC-based Products by World Areas, Annual 1982,* FT 210. Washington, D.C.: Government Printing Office, May.

U.S. Department of Commerce (1977a) *Census of Manufactures 1977, Subject Series, Fuels and Electric Energy Consumed,* table 3, pp. 4-26 and 4-29. Washington, D.C.: Government Printing Office.

U.S. Department of Commerce (1977b) *Census of Manufactures 1977.* Washington, D.C.: Government Printing Office.

U.S. Department of Commerce (1980) *Annual Survey of Manufactures, 1980.* Washington, D.C.: Government Printing Office.

U.S. Department of Labor, Bureau of Labor Statistics (1983) *Productivity Measures for Selected Industries, 1954-82,* Bulletin 2189. Washington, D.C.: Government Printing Office, December.

Vernon, Raymond (1966) "International Investment and International Trade in the Product Cycle," *Quarterly Journal of Economics* 80, 2, pp. 190-207.

Wheaton, William (1983) "Interstate Differences in the Level of Business Taxation," *National Tax Journal* 36, 1 (March), pp.83-94.

5

Local Employment Growth
and the
Reemployment Success
of Displaced Workers

Chapters 3 and 4 demonstrate that the numbers of displaced workers are larger in the slow-growth industrialized states of the North, but the rate of labor displacement for two of the three industries and the rate of plant closures for all three industries are even across the regions. Workers in the metalworking machinery and electronic components industries located in the healthy economies of the Sun Belt are as likely to be displaced by a plant closing as workers in the slow-growth economies of the Rust Belt. These findings suggest that labor displacement may be a nationwide problem.

If manufacturing workers displaced in strong labor markets are readily reabsorbed into the workforce, then assistance for dislocated workers should target depressed communities. If, however, displaced workers in strong, as well as weak, local economies face difficult adjustments and large financial losses after displacement, a nationwide strategy is appropriate. This issue is the focus of chapter 5.

Three questions are addressed. (1) What impact does local labor market growth have on the readjustment of displaced manufacturing workers? (2) Are there specific demographic groups who experience large financial losses after displacement, even when laid off in growing labor markets? (3) What impact does worker prenotification have on the reemployment success of displaced manufacturing workers?

As part of their Current Population Survey in January 1984, the Bureau of Labor Statistics identified displaced workers for the first time. We used this data base to explore the above issues. There are five findings. (1) The reemployment success of displaced manufacturing workers is more sensitive to local conditions in the workers' industry than to conditions in the total local economy. (2) Workers displaced from locally declining industries experience the longest stretches of unemployment and the greatest declines in living standards. (3) All middle-age, blue-collar workers with a stable work history, even those displaced in strong local labor markets, experience large financial losses after displacement. (4) Displaced workers' reemployment success is influenced not only by local labor market conditions, but by the number of years worked prior to layoff, the state of the national economy in the year of the layoff, and the workers' sex, race, and occupation. (5) Workers who are prenotified of a closing are not, as a group, more successfully reemployed than those who are not prenotified.

The remainder of this chapter is divided into six sections. The first describes the Current Population Survey—Supplement on Displaced Workers. The second section reviews prior literature on the effect of labor market growth and worker demographics on reemployment. The third section defines a measure of the financial costs of displacement and the study methodology. The fourth section presents regression results. The fifth section discusses the role of prenotification on workers' reemployment success, and the final section summarizes findings.

Current Population Survey—Supplement on Displaced Workers

In January 1984, as a part of the Current Population Survey (CPS), the Bureau of Labor Statistics (BLS) identified workers who lost their jobs after January 1, 1979 as a result of a plant closing, the phase-out of a shift, the closure of a self-operated business, the completion of seasonal work, or slack work. While there is no widely accepted definition of displacement, for purposes of this study we define displaced workers as those who lost their jobs for the first three reasons given above. This definition was chosen because it is unlikely that a worker laid off because of a plant closing, the phase-out of a shift, or the failure of a self-operated business will be rehired at his or her old job, whereas workers displaced for the last two reasons may be reemployed by the same firm. Among those surveyed, 82 percent lost their job as the result of a plant closing or relocation, 15 percent as the result of the phase-out of a shift, and 3 percent as the result of a failure of a self-operated business. We also define a displaced worker as one who completed at least one full year on the job prior to layoff, since much of the current proposed plant closing legislation would require that workers be employed at least one year. Based on this definition, the BLS-CPS data indicate that slightly more than 3 million workers were displaced between January 1979 and December of 1982.

The January 1984 CPS provides information on the length of the displaced worker's unemployment spell, and his or her previous and current wage, old and new industry affiliation, age, years at the firm prior to layoff, education level, sex, race, and whether he or she was prenotified or expected the closing, along with other information. We supplemented these data with figures on employment gains and losses from 1979 to 1983 in the worker's labor market and on average weekly unemployment insurance payments in 1980 by state.

Shortcomings in the Data

There are a number of shortcomings in the data for the purposes of this study, One is that many workers, especially those displaced close to the 1984 survey date, did not complete their unemployment spell prior to the survey, making it impossible to determine the length of their unemployment period and their subsequent wage. A failure to correct for this truncated sample is likely to result in (1) underestimating the duration of the average unemployment spell, (2) overestimating the average postdisplacement wages, and (3) making faulty comparisons across workers. For example, a worker displaced in December of 1983 and still unemployed in January of 1984 would register four weeks of unemployment. A worker displaced earlier in the year, who was unemployed for four weeks and then reemployed would register the same unemployment spell. The two unemployment spells should not be treated as equivalent. In addition, the reemployment wage among those displaced close to the survey date will be biased upwards. The most successful workers are most likely to be reemployed quickly, prior to the survey date, and these workers probably earn higher-than-average postlayoff wages.

In order to insure that most workers in our sample had a chance to complete their unemployment spell prior to the January 1984 survey date, our analysis excludes all workers laid off in 1983 and 1984. The years to be excluded were determined by examining the distribution of unemployment spells for each year of layoff. These distributions were done separately for each year because the length of the spells could vary by year, with longer spells more likely in years of a weak national economy. Unfortunately, the data do not include information on the month a worker was laid off. To be conservative, we assumed all workers were laid off in December of the layoff year. If all workers laid off in 1982 were laid off in December of 1982, 75 percent would have completed their unemployment spell by the January 1984 survey date. Thus, these observations are included in the study. If workers displaced in 1983 lost their jobs in December of that year, only 40 percent would have completed their unemployment spell as of the survey date; thus, these observations are excluded.

A second shortcoming of the data set is the exclusion of nonmetropolitan observations. The only information contained in the CPS on a worker's geographical location is SMSA and state of residence. States are too large to be used as a proxy for a labor market, thus the analysis is limited to SMSA residents.

Another possible shortcoming is that the CPS does not report the location of layoff, although there is a record as to whether the worker moved. All movers are assigned to the city of residence at the time of the survey. Approximately 19 percent of displaced manufacturing workers in SMSAs relocated from another county or city, so the potential for bias is substantial. Presumably, most migrants moved from declining areas to growing SMSAs, most likely after a prolonged and discouraging job search in the declining area. The resulting bias would make conditions look more favorable in declining areas and less favorable in growing SMSAs than is actually the case. To test for this possibility, the results were estimated with and without movers. The direction of the bias is as expected, but, as explained below, not large.

A fourth problem is recall bias. Horvath (1982) found that when individuals are asked to recall the length of unemployment spells that occurred 6 to 12 months in the past, they tend to underestimate the length of the spell. This suggests there is an even greater potential bias here, where workers are asked to recall the amount of time they were unemployed up to four years earlier.

Previous Literature

Effects of Labor Market Conditions on Reemployment

Prior to the release of the January 1984 CPS, most empirical work on the experience of displaced workers was based on case studies (see Gordus, Jarley, and Ferman 1981). While pre- and postclosing surveys from a particular plant can illuminate many of the consequences of a shutdown, it is unclear to what extent the results can be generalized to all displaced workers. This is particularly true in the plant closing literature, where most cases have focused on closings in declining and manufacturing-dependent communities. These studies have shed a great deal of light on the impact of workers' age, education, race and sex on their reemployment success, but can tell us very little about the adjustments for these demographic groups in expanding local economies.

Case studies that do address the impact of labor market conditions on reemployment are those of the Armour and Company plant closings. The results from studies of five Armour and Company meat packing plant closures show the condition of the labor market in which the closings occurred had a crucial effect on the success of training and transfer programs (Ullman 1969). In the 1960s, Armour and Company, the United Packing Workers Union, and the Amalgamated Meat Cutters and Workman's Union agreed to assist workers who were displaced as a consequence of Armour closings. Offices were set up in Forth Worth in 1962, Sioux City in 1963, Kansas City in 1964, Peoria in 1967, and Omaha in 1968 to coordinate and establish counseling services, offer retraining and placement programs, and provide job search support groups. Qualitative evaluations conducted after the five experiences indicated that the strength of the local labor market had a crucial effect on the placement records for each city (see also Shultz 1964; Conant 1965; and Stern 1969).

In a study of displaced steel workers, Jacobson (1977), found that if the worker was displaced into a local labor market where the unemployment rate was 1.4 percentage points above the city's average rate, financial losses over the first six years after displacement were about 8 percentage points higher than average. He also found that younger workers' losses were particularly sensitive to the unemployment rate. Younger workers experienced negligible losses in strong labor markets and substantial losses in weak local markets.

A third study, by Bendick and Devine (1981), found 14.8 percent of workers displaced in declining SMSAs were unemployed for more than 26 weeks, while only 7.6 percent of workers displaced in growing SMSAs were unemployed for more than 26 weeks. The authors defined displaced workers as those who were unemployed for longer than eight weeks, actively seeking employment, and between the ages of 22 and 54.

Podgursky and Swaim (1987), using the same CPS data used here, found high area unemployment levels, measured as SMSA rates for workers located in the largest SMSAs and state rates for all other workers, reduced the reemployment earnings of displaced workers. Unemployment was measured as the average rate between the year of displacement and 1984. For every 1 percent increase in the unemployment rate, postdisplacement earnings fell by between 1 and 2 percent.

Effects of Worker Characteristics on Reemployment

A large number of case studies have focused on the effects of a worker's age, education level, race, and sex on the postdisplacement adjustment.

Age
Mature workers are expected to encounter larger wage losses than younger workers because they have invested in job-specific human capital that was rewarded prior to layoff but may not be transferable to a new

job. Age discrimination, employers' reluctance to spend resources on retraining and pension plans for someone with a relatively short worklife, and the actual or perceived inflexibility of older workers are additional reentry barriers for older workers. This hypothesis is supported by a number of studies.

Hammerman (1964) conducted case studies of five plant closings and reported that in four of the five cases, workers over 45 years of age had significantly higher unemployment rates than those below that age. When older workers did obtain jobs, they experienced greater losses in hourly earnings than did younger workers. Foltman (1968), in a study of the Wickwire plant closing near Buffalo, New York, found older workers had greater difficulty in obtaining jobs. Nine months after the shutdown, 83 percent of the 20-29 age bracket, 92 percent of the 30-39 age bracket, 73 percent of the 40-49 age bracket, and only 44 percent of the over 50 age bracket had found new jobs. After controlling for education and family status, Dorsey (1967) found that age was an important variable in explaining the unemployment spell of workers laid off from a Mack Truck plant in Plainfield, New Jersey. Every year of age resulted in an average of .2 additional weeks of unemployment.

Jacobson and Thomason (1979), Holen, Jehn, and Trost (1981) and Podgursky and Swaim (1987) conducted cross-industry and cross-region studies of worker displacement and also found older workers experience larger postdisplacement losses than younger workers.

Education

Levels of formal education are also expected to affect a worker's postlayoff experience. More highly educated workers should achieve higher wages at their postdisplacement job and shorter unemployment spells than their less-educated counterparts. Workers with more education should conduct more effective job searches, make better presentations at an interview, be more ambitious or assertive, and be perceived by potential employers to be more flexible and trainable.

The evidence of a relationship between length of unemployment and education is mixed, but generally shows duration decreases as education increases. In his case studies of five plant closings, Hammerman (1964) found that 6-21 months after shutdown, high school graduates experienced lower unemployment rates than those who had not completed high school. The least educated workers also took the sharpest cuts in wages. In the study of the Wickwire plant closing in Buffalo, Foltman (1968) found that nine months after the closing only 49 percent of those who had not completed high school had found jobs. Of those who had completed high school, 79 percent had obtained jobs; of those who had some college education, 82 percent had found jobs. Foltman also reports that workers who were more educated were less likely to accept lower incomes when reemployed. Dorsey (1967), in the Mack Truck study mentioned above, found that better-educated workers experienced shorter unemployment spells after layoff, and a survey of the reemployment of workers from the Playskool plant in Chicago found higher placement rates for workers with high school diplomas (O'Connell 1985).

Race

A number of plant closing case studies report that minorities do relatively poorly in terms of reemployment after a permanent layoff. Stern (1971), in a study of displaced meat packers in Kansas City, found blacks made $204 less per year than whites when reemployed. Palen and Fahey (1968), in a study of the Studebaker plant closing in South Bend, Indiana, found 39.9 percent of whites and 60 percent of non-whites were unemployed four months after the shutdown. Aiken, Ferman, and Sheppard (1968) found a similar result in their study of the Packard plant closing.

Sex

Similar findings are reported for women. Hammerman (1964), in his study mentioned above, noted female unemployment rates three times the rate of males; Stern (1971) reports in his study of displaced meatpackers that females made $2,123 less per year than males when reemployed; and Lipsky (1970) found that women displaced during a General Mills shutdown were unemployed an average 24.8 weeks, compared to only 16.5 weeks for the males. Only Lipsky's study controls for workers' age, race, education and previous income. A similar result for women is reported by Jacobson and Thomason (1979), Holen, Jehn, and Trost (1981), and Podgursky and Swaim (1987) in their studies of the impact of displacement on earnings losses.

Level of Unemployment Insurance Benefits

Numerous empirical studies outside the plant closing literature have found evidence that individuals who receive higher unemployment insurance (UI) benefits prolong the job search (Classen 1977; Feldstein 1974; and Welch 1977). Consequently, in addition to labor market conditions, and workers' age, education level, race and sex, we expect the size of a state's unemployment insurance payments to influence the length of the unemployment spell and the reemployment wage of displaced workers.

Methodology

Estimating Financial Losses From Displacement

A displaced manufacturing worker experiences financial losses as a result of the unemployment spell and any wage cuts upon reemployment. We measured these financial losses as the present value of the predisplacement wage five years into the future subtracted from the present value of zero wages during the unemployment spell plus the postdisplacement wage thereafter, continuing five years into the future.

$$(1) \quad CD = \sum_{i=s+1}^{260} \frac{cw}{(1+r)^i} - \sum_{i=1}^{260} \frac{ow}{(1+r)^i}$$

where s = Weeks unemployed.

r = Annual discount rate of 10 percent.

ow = Previous weekly wage.

cw = Current weekly wage.

CD = Cost of displacement.

i = Weeks (i=0 in week of job loss).

A 10 percent discount rate was chosen to reflect an average annual rate of inflation of 6.5 percent over the period 1979-1987, and a marginal rate of substitution between present and future consumption of 3.5 percent (see Tresch 1981, p. 489). Hereafter, this variable is referred to as the cost of displacement (CD). A negative value signifies financial losses. A positive value indicates a financial gain.

The advantage of this measure of the financial costs of displacement is that it summarizes, in one indicator, the payoff to a worker who trades a longer unemployment spell for a higher-paying job. Its disadvantage is that its accuracy depends on the assumption a worker's prelayoff wage would have continued five years into the future, and the reemployment wage continues unchanged for up to five years (after the unemployment spell). These measurement errors probably result in an underestimation of both the present value of the reemployment wage and prelayoff wage, with the largest underestimates for younger, well-educated workers displaced in growing regions. This is the group most likely to receive wage increases with time. Figure 5.1 compares actual losses with those measured in this study. The gap between actual losses and measured losses depends upon the slopes of the earning profile, which are, unfortunately, unknown.[1]

We also assume the 17 percent of the sample unemployed on the survey date will be unemployed for the whole period. For both unemployed and part-time workers, we assume that the value of their leisure is zero, which may bias financial losses upwards. An additional shortcoming is that the measured financial losses ignore the value of fringe benefits. However, this may be more than balanced by the fact that we do not include the value of transfers to unemployed workers.[2]

The effect of local labor market conditions and worker demographics on the financial costs of displacement are tested with the following model. The variables are defined in table 5.1.

$$(2) \quad CD = \lambda_0 + \lambda_1 AE_s + \lambda_2 IE_s + \lambda_3 YRLO_i + \lambda_4 YRWKD_i$$
$$+ \lambda_5 Age_i + \lambda_6 ED_i + \lambda_7 Race_i + \lambda_8 Sex_i + \lambda_9 BEN_s + \lambda_{10} OCC_i.$$

Similar models are also tested, replacing CD with the change in nominal weekly wages (CWW) and the number of weeks unemployed (WKS). The latter two models offer insight into the reasons for financial losses. Are losses, for example, due to long unemployment spells or large losses in nominal weekly wages?

Figure 5.1
Comparison of Measured Financial Losses with Actual Losses

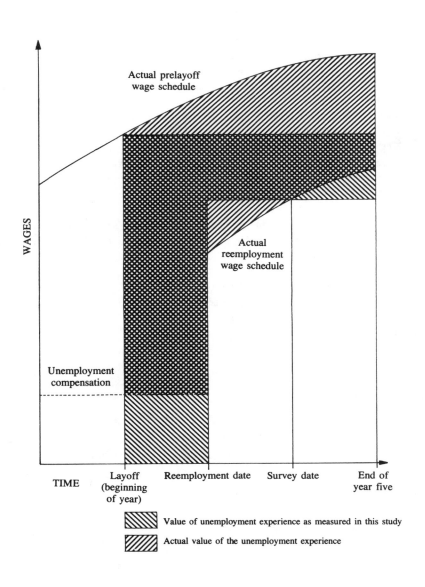

Table 5.1
Variables Included in the Model
(Equation 2)

Conditions in the Local Labor Market

AE = Annual average net employment change for all industries January 1979 to December 1982 in thousands of jobs.

IE = Annual average net employment change, in 2-digit SIC level industry of displacement, 1980 to 1983 in thousands of jobs.

YRLO = Dummy variable representing the state of the national economy in the year laid off. Value=1 if recession year 1980 or 1982 and 0 if an expansionary year 1979 or 1981.

Characteristics of the Worker

YRWKD = Years worked at that job prior to layoff.

Age = Worker's age in January 1984.

ED = Worker's years of formal education in January 1984.

Race = Dummy variable with 1=white and 0=nonwhite.

Sex = Dummy variable with 1=female and 0=male.

BEN = Average weekly unemployment insurance payment in a worker's state.

OCC = Dummy variable representing the worker's prelayoff occupation. Value=1 if a blue-collar worker and =0 if a white-collar worker.

Dependent Variables

CD = Total cost of displacement (equation 1).

CWW = Change in nominal weekly wages.

WKS = Weeks unemployed.

Subscripts

i = Individual.

s = Local labor market.

The coefficients on YRLO, YRWKD, Age, Sex and OCC are expected to be negative, and the coefficients on AE, IE, ED, and Race are expected to be positive. *A priori,* the sign on the coefficient for BEN is ambiguous. Strong empirical evidence exists to show that individuals who receive UI compensation extend their job search, reducing the present value of the unemployment experience. However, the expectation is that this extended job search will allow individuals to find employment more suited to their skills and thus to increase their reemployment wage beyond what it would have been if the worker had been forced by economic necessity to take the first job that came along. UI benefits should therefore extend the weeks unemployed and increase the new wage. But it is unclear whether the reemployment wage, extended five years into the future, will be large enough to offset the losses from a prolonged spell of unemployment.

Findings

The mean costs of displacement, the change in nominal weekly wages, and weeks unemployed are shown in table 5.2 for all displaced manufacturing workers, and for comparison purposes, all displaced workers.

Table 5.2
Unemployment and Reemployment Statistics in January 1984
for Workers Displaced Between January 1979 and December 1982
All Industries and All Manufacturing

Cost of displacement[a] $	Mean change in nominal weekly wages[b] (% gain or loss)	Employed[c] %	Unemployed[c] %	Out of labor force[c] %	Weeks unemployed[d] %
All workers					
−16,543.	4.03 (1.0%)	68	15	17	32
All manufacturing workers					
−21,531.	−12.01 (−4.1%)	67	16	17	37

a. Universe includes all workers in labor force at time of survey, except self employed.

b. Universe includes only wage and salaried workers. Excludes self employed. Equal to current nominal weekly wage minus previous nominal weekly wage.

c. Universe includes all workers.

d. Universe includes all workers in labor force at time of survey, except self employed.

Displaced workers who found new jobs did so at an average increase of $4 per week in wages. During the same period, the average weekly wage for all workers in the U.S. increased by an average of $17.50 per year, or $11.00 over a 7.5-month period, the average unemployment spell for displaced workers. Therefore, displaced workers sustained a net loss of $7 per week when compared to earnings for all workers over a comparable time period.

Displaced manufacturing workers suffered larger financial losses and longer unemployment spells than the average displaced worker. The average displaced manufacturing worker lost $12 per week in wages when reemployed. This drop occurred during the same period in which the average nominal weekly manufacturing wage increased by approximately $24.00 per annum (U.S. Department of Commerce 1985, p. 417).

We now turn to the question of how displaced workers fare in growing versus declining labor markets. Table 5.3 compares the costs of displacement in declining and growing labor markets for two industries, machinery and transportation manufacturing, and for all manufacturing. Machinery and transportation manufacturing were selected for disaggregation because they are the only 2-digit SIC code industries in the survey with sufficient numbers of observations to yield meaningful results. The table shows that the costs of displacement are sensitive to local industry employment growth. For machinery manufacturing, workers displaced in growing areas experienced, on the average, a net loss of $1,173 in the five years following displacement, compared to a net loss of approximately $17,182 for workers displaced in declining labor markets. Displaced transportation workers and all manufacturing workers also experienced greater financial losses in declining than in growing labor markets.

Table 5.3 also compares the average change in nominal wages, the current employment status, and the average length of the unemployment spell in growing and declining local labor markets. All indicators are sensitive to local labor market conditions. However, for all manufacturing, the drop in wages for reemployed displaced workers is nearly equivalent in both growing and declining SMSAs.

Table 5.3

Comparison of the Unemployment and Reemployment Experience of Displaced Manufacturing Workers in Growing and Declining Labor Markets, for Selected Industries and All Manufacturing

Mean cost of displacement[a] $	Mean change in nominal weekly wage[b] $ (% change)	Current Status[c]			
		Percent working	Percent unemployed	Percent out of labor force	Mean weeks[d] unemployed
Machinery Manufacturing (SIC 35)					
Growing labor markets					
−1,173	45.44 (+12.7)	82	12	6	21
Declining labor markets					
−17,182	−49.96 (−12.7)	68	20	12	31
Transportation Manufacturing (SIC 37)					
Growing labor markets					
−6,130	−9.49 (−2.5)	94	0	6	10
Declining labor markets					
−45,942	−115.27 (−29.9)	65	23	12	67
All Manufacturing					
Growing labor markets					
−19,758	−10.97 (−3.2)	68	16	16	31
Declining labor markets					
−31,298	−12.13 (−3.4)	58	23	19	43

SOURCE: U.S. Bureau of Labor Statistics, Current Population Survey, Supplement on Displaced Workers, January 1984.

NOTE: For 2-digit SIC industries, growing labor markets defined as SMSAs with net job gain of 1450 jobs or more in that industry; declining labor markets as SMSAs with net job loss greater than 49 jobs. For all manufacturing, growing labor markets defined as net manufacturing employment growth equal to or greater than 9,500 employees and declining labor markets as SMSAs with net job loss greater than 49 manufacturing jobs. Selected industries with weighted totals of 10,000 population in both growing and declining SMSAs.

a. Includes all displaced workers currently in the labor force. Excludes self-employed. Value as calculated in equation (1).

b. Includes only currently working wage and salaried workers. Excludes self-employed. Equal to current weekly wage minus previous weekly wage. A positive value indicates a gain in nominal wages.

c. Includes all displaced workers, as defined in text. Definitions for each category are as defined by the U.S. Bureau of the Census in its Current Population Survey.

d. Includes all displaced workers currently in the labor force. Excludes self-employed.

Impact of Local Labor Market Conditions
and Worker Characteristics

The results of equation (2) are shown in table 5.4. The dependent variable in column I is the cost of displacement. (Again, a negative value indicates a loss.) The dependent variables in columns II and III are the change in nominal weekly wages and number of weeks unemployed, respectively.

Several findings are of interest. First, the reemployment success of displaced manufacturing workers is more sensitive to local growth in their industry of displacement (IE) than in total local employment (AE). One interpretation is that manufacturing workers do not move easily into new industries, and when they do so it is at a substantial loss in income. The results on the local industry growth coefficient can be interpreted as, for every 1,000 industry jobs added to the local labor market in a year, the losses experienced by a displaced worker are reduced by $4209. The coefficient on total employment growth (AE) is both relatively small and statistically insignificant.

We expected that the inclusion of movers in their SMSA of current residence rather than area of displacement would bias the coefficients on local economic growth downwards because some of the most hard hit and discouraged workers displaced in declining areas moved to growth economies. Reestimating equation (2) without movers, we found that the direction of the bias was as expected, but not large. Without movers in the equation, the coefficient on total employment (AE) is $19 and on local industry growth (IE) is equal to $4226.

While there is no evidence that growth in total local employment reduces the loss in weekly wages, there is evidence that workers displaced in expanding local economies are unemployed for shorter periods of time than workers in weak economies (see column III). However, the impact is small. The coefficient on AE in column III can be interpreted to mean that a net increase of 1000 local jobs in all sectors reduces the unemployment spell by 2 percent of a week (or less than one hour). In contrast, an additional 1000 jobs in the workers' industry of displacement reduces the unemployment spell by 2-1/2 weeks. The large financial losses of workers displaced where their industry is declining is due to both declines in weekly wages and lengthier periods of unemployment (columns II and III).

Table 5.4
The Impact of Labor Market Conditions and Worker Demographics on the Cost of Displacement, Ordinary Least Squares

Dependent variable	I Cost of displacement	II Change in nominal weekly wage	III Weeks unemployed
Constant	−28033.08 (−.88)	−111.83 (−.71)	59.15* (2.57)
AE	12.82 (1.11)	.04 (.69)	−.02* (−2.25)
IE	4208.56* (2.17)	20.08* (2.04)	−2.57** (−1.88)
YRLO	−12555.63* (−2.11)	−70.34* (−2.34)	−9.84* (−2.31)
YRWKD	−1285.01* (−2.48)	−6.47* (−2.47)	.64** (1.77)
Age	−292.79 (−.94)	−1.46 (−.92)	.08 (.36)
ED	1455.61 (1.18)	8.81 (1.43)	−.73 (−.88)
SEX	17893.91* (2.75)	78.45* (2.38)	.51 (.11)
Race	14134.98* (1.96)	58.63 (1.61)	−17.71* (−3.35)
BEN	39.83 (.16)	.17 (.13)	−.09 (−.48)
OCC	−10251.15 (−1.48)	−44.01 (−1.27)	15.83* (3.16)
F	5.10*	4.56*	6.22*
R^2	.22	.20	.23
No. of observations	186	188	213

* Statistically significant at 5 percent level of confidence.
** Statistically significant at 10 percent level of confidence.
( ) t-statistics.

As expected, workers laid off when the national economy was in decline experienced larger financial losses, due to greater declines in their reemployment wage. Contrary to expectation, however, workers displaced during the recessions experienced shorter periods of unemployment by nearly 10 weeks. One explanation is that workers displaced when the national economy is weak lower their reservation wage sooner and, therefore, find jobs more quickly than workers displaced in a stronger national economy.

Workers who had committed more years to their employer prior to layoff (YRWKD) experienced larger financial losses than recently employed workers. This finding indicates that earnings of senior workers reflected union rates or firm-specific skills of less value to new than prior employers. The results show that for every year of employment at their prelayoff job, workers lost $1285 in total income and $6.47 in weekly wages once reemployed. When YRWKD is added to the equation, the coefficient on Age becomes insignificant. The two variables are highly correlated, with a Pearson correlation coefficient of .56 (see appendix 5.2).

Surprisingly, the coefficient on education is insignificant. Again, this may be due to multicollinearity among the independent variables. Occupation and education are correlated (the correlation coefficient = .40) and when occupation is dropped, education is statistically significant.

In contrast to the numerous case study findings, this analysis shows women do better than men in recouping their prelayoff wage (but do not experience shorter unemployment spells). There are two possible explanations. First, women probably earned lower wages in the manufacturing sector than men, and therefore had less difficulty recovering their prelayoff wage. Second, women are likely to hold service sector-type jobs in manufacturing, and thus may be able to put those skills to use in another industry, including the growing service sector. Men, on the other hand, may have been trained for jobs specific to their industry or firm. When a similar job is not available in another firm, they may have to again start at the bottom of the wage ladder. Our findings on the reemployment pattern for women may differ from earlier studies because previous studies explained wage levels rather than wage changes.

In support of case study and other statistical studies, nonwhites appear to experience larger financial losses than whites, by $14,135. These losses are due to blacks' longer unemployment spells, by nearly 18 weeks, and not to larger losses in the reemployment wage of blacks.

The generosity of state unemployment insurance (UI) payments has no impact on the cost of displacement, the reemployment wage, or the weeks unemployed. One explanation for this finding is that displaced workers behave differently from all unemployed, and do not prolong their job search where UI benefits are larger. A second possibility is that average weekly UI benefits may be a poor measure of a worker's actual compensation. Displaced workers, in addition to UI benefits, may receive Trade Adjustment Assistance, extended health care benefits, severance payments from the company, or compensation from their union. Unfortunately, the BLS survey did not ask respondents about supplemental compensation.

Finally, there is evidence that blue-collar workers are unemployed for an average of 15.8 weeks longer than white-collar workers, but little support for the hypothesis that blue-collar workers are less likely than white-collar workers to recoup their prelayoff wage. To summarize, we find the reemployment success of displaced workers is more sensitive to local employment growth in their industry of displacement than to growth in all industries, and that black, blue-collar workers with long commitments to their prelayoff employer experience the largest financial losses.

Impact of Prenotification on the Reemployment Success of Displaced Workers

The CPS data base also provides information on whether displaced workers were notified by employers their plant would close. Among the numerous federal proposals to deal with worker displacement, employee prenotification of an impending closing has the most support from both labor and Congress. In 1985, proponents of a national plant closing bill put their strength behind legislation requiring only prenotification, ignoring a host of other issues such as severance payments, extended health care, mandatory negotiation over whether the closing can be averted, and employee retraining. This legislation was narrowly defeated in a House vote on November 21, 1985, 203 to 208. A similar bill, requiring 60 days notice for firms with 100 employees or more, was revived in the 100th Congress. The bill was passed by both Houses, but vetoed by President Reagan in mid-1988. The House of Representatives overrode the presidential veto. A majority in the Senate supported passage, but they could not muster a two-thirds vote. The legislation was a omnibus bill, including a wide range of trade related initiatives. However, the mandatory prenotification clause attracted the most controversy and was responsible for the veto of the total trade package.

Maine and Wisconsin currently require advance notice, although the legislation is only weakly enforced; and during the period 1975-83, 30 states introduced more than 125 bills relating to plant closings. More than 90 percent of these bills had provisions requiring advance notice of shutdowns (Ehrenberg 1985).

The legislative emphasis on prenotification is motivated by the belief that adjustment problems are particularly acute where workers and communities receive little advance notice of a closing, and consequently cannot plan for an employee buyout, reuse of the physical facility, or a period of personal or community fiscal stringency. Advance notice also gives employers, workers and the community a chance to explore ways to keep the plant operating through, for example, the identification of a new owner. Finally, prenotification gives local governments a chance to set up retraining and counseling programs for displaced workers. It is difficult to reach displaced workers once the plant has closed and the workers have dispersed.

Opponents of federal- or state-mandated prenotification claim such legislation would have adverse effects on the firm and the community. Potential adverse effects on the firm include the departure of the firm's most productive workers prior to the closing, falling stock prices for the plant's parent, and precedent for further government involvement in the affairs of business. Opponents also claim that mandatory prenotification would inhibit local economic growth by discouraging firms from locating or expanding in states where an advance notice law is in effect (McKenzie 1982).

The evidence on the frequency of employee prenotification indicates that, in the absence of legislation, little or no prenotification is common. The following story by a displaced worker from a branch of Dresser Industries is typical of the anecdotal evidence.

> When Dresser got us down to about 50 people, they told us we were in trouble. They said we'd have to give up something. It wasn't contract time until June of '84 and we had no reopener clause, so those of us that were left gave the company a full week without pay.
>
> We were worried and scared, too, I guess.
>
> We asked management if the plant was going to close. At first, we were told no, and later were told, if the plant was going to shut down, we would be given plenty of advance notice.
>
> But that isn't the way it worked. In October we were told the plant was going to be shut down within three weeks. And on November 15, 1982 it shut down and closed.[3]

In 1980, only about 13 percent of all major collective bargaining contracts included prenotification clauses. About one-half of those did not specify the length of advance notice and only 34 percent agreed to an advance notice of 31 days or more (Wendling 1984). Although these figures indicate that a minority of employers are required to give advance warning of a closing, the figures may actually overstate the number of protected workers. Employees not covered by a collective bargaining agreement are even less likely to have a legal right to advance notice.

A second study on the frequency of prenotification indicates higher rates of prenotification. Gainer (1986) surveyed 355 firms that closed plants and found that 20 percent of the establishments had provided their workers with over 90 days of advance notice and 40 percent had provided less than 14 days or no advance notice. The higher prenotification figures in the Gainer study could be because (1) many firms who are not required to give prenotification do so; (2) the Gainer study is more recent and, with increased publicity, more firms are giving prenotification; or (3) Gainer interviewed larger firms (firms with 100 employees or more), which are more likely than the average size firm to give advance notice.

One study of the impact of prenotification on reemployment success is by Folbre, Leighton, and Roderick (1984). They used data from Maine, prior to the passage of the 1982 mandatory advance notice law, and found that unemployment rates in local communities where a closing occurred did not rise as high and fell faster where employees had been given advance notice of the closing. The authors argue that this evidence "strongly suggests that Maine workers significantly benefitted from advance notification of job loss" (p. 195).

Holen, Jehn, and Trost (1981), using a sample of 9500 workers from 42 plants that closed, found a statistically significant and positive relationship between advance notice and postdisplacement losses. In other words, workers given advance notice experienced greater postdisplacement losses than those who were not given advance warning. The authors speculated that "advance notice may be given in just those cases where losses are expected to be high" (p. 27). Although the authors suggest the variable prenotification may be endogenously determined, they do not make the appropriate corrections in their empirical work.

The BLS's Survey of Displaced Workers permits the exploration of several of the issues posed above, including whether a displaced worker was prenotified, the extent to which prenotification encourages preclosing departures, and the impact of prenotification on workers' reemployment success.

The question posed to workers was, "Did you expect a layoff or had you received advance notice of a layoff or a plant or business closing?" The survey includes no additional information as to the length of the advance notice, or whether the worker was informed by management or expected the plant to close because of such factors as rumors of poor profit performance. Even given these shortcomings, there are several interesting findings.

Among all displaced manufacturing workers living in SMSAs at the time of the survey, 61 percent said they were either given advance notice or expected their plant to close. Among the 61 percent who expected a closing, only 9 percent left their jobs before they would have been laid off.[4]

The impact of prenotification on displacement losses was tested using both the ordinary least squares estimation of equation (3) below and two-stage least squares (2SLS):

(3) $CD = \alpha_0 + \alpha_1 AE_s + \alpha_2 IE_s + \alpha_3 YRLO_i + \alpha_4 YRWKD_i$
$+ \alpha_5 Age_i + \alpha_6 ED_i + \alpha_7 Race_i + \alpha_8 Sex_i + \alpha_9 BEN_s$
$+ \alpha_{10} OCC_i + \alpha_{11} PN$

Where the variables are defined in table 5.1, except:

PN = Dummy variable, with 1 = the worker was prenotified or expected the closing and 0 = the worker was caught by surprise.

Column I in table 5.5 reports the results, which indicate larger losses for prenotified than non-notified workers. This result is similar to that of Holen, Jehn, and Trost (1981) who used both a different data set and different measures of income losses. One explanation for this counterintuitive result is that, for humanitarian reasons, employers are more likely to prenotify workers who will experience large losses. Another reason is that employers may be more likely to prenotify workers in weak labor markets, where losses will be large, because employees will have relatively few employment options. Employers may know the probability is small that these workers will leave prior to the closing.

If any of the above explanations are true, there will be simultaneity bias in the estimated coefficients. If prenotification is a function of expected financial losses, the assumption of independence between the independent variables and the error term is violated (see Maddala 1977, pp. 231-235).

The 2SLS method of estimating simultaneous equations was used to correct for this problem. The simultaneous equation system includes equations (3) and (4), where equation (4) explains the probability a worker was prenotified about the closing. A predicted value of prenotification ($\widehat{PN}$) was generated using the reduced form of equation (4). The variable PN in equation (3) was then replaced with $\widehat{PN}$.

(4) $\widehat{PN} = \gamma_0 + \gamma_1 CON_j + \gamma_2 UN_j + \gamma_3 ME_s + \gamma_4 S_j + \gamma_5 CD_i$

where

CON = Percent of output produced by the four largest firms in 1980.

UN=Dummy variable =1 if the SMSA and industry had above
median levels of unionization and 0 if they did not.

ME=Annual average change in manufacturing employment, 1979
to 1982.

S=Average size of plant in 1980.

CD=Costs of displacement.

j=Industry of displacement at the 3-digit Standard Industrial Code
level.

s=Local economy.

i=Individual.

The above independent variables were included because it was expected workers in concentrated industries would be more likely to
negotiate advance warning clauses in labor contracts; unionized workers
would be more likely to be prenotified for the same reason; workers
in growing economies would be less likely to be prenotified because
employers feared they would leave early; and workers in large plants
were more likely to suspect or hear rumors about an impending closure.
The reduced form of equation (4) was estimated with a logit model,
since the dependent variable is dichotomous. $\hat{PN}$, the probability a
worker was prenotified is, however, a continuous variable.

The results of the 2SLS estimation are reported in columns II, III,
and IV of table 5.5. While the results show the expected sign on $\hat{PN}$,
the coefficient is still not significant. There is no evidence prenotified
workers experience smaller financial losses than non-notified workers.

One hypothesis is that prenotification only reduces financial losses
when workers are located in growing labor markets. This hypothesis
was tested by including interaction terms between prenotification and
local employment growth. Again, neither interaction term is significant, reinforcing evidence that prenotification does not reduce losses
to displaced workers (see table 5.6).

The only workers who appear to benefit from advance notice are those
who quit prior to layoff (see table 5.7). Table 5.7 reports the results
when only prenotified workers are included in the analysis and the dummy variable PN is replaced with a dummy on prenotification and early
quit (PN&QT). The variable PN&QT is equal to 1 when a prenotified

Table 5.5
Impact of Prenotification on the Reemployment Success
of Displaced Manufacturing Workers

Dependent variable	I Ordinary least squares Costs of displacement	II Costs of displacement	III Two stage least squares Change in nominal weekly wage	IV Weeks unemployed
Constant	−25502.95	−57391.72	−269.31	59.64**
	(−.74)	(−1.28)	(−1.20)	(1.80)
PN	−8039.32	68168.21	360.91	−22.89
	(−1.30)	(1.10)	(1.15)	(−.49)
AE	11.61	14.73	.05	−.02*
	(1.00)	(1.22)	(.83)	(−2.03)
IE	3861.17*	5734.21*	28.20*	2.95*
	(1.98)	(2.49)	(2.43)	(−1.77)
YRLO	−11512.52**	−19963.67*	−109.05*	−6.77
	(−1.92)	(−2.33)	(−2.52)	(−1.07)
YRWKD	−1267.87*	−1525.10*	−7.77*	.65**
	(−2.45)	(−2.87)	(−2.89)	(1.75)
Age	−276.21	−265.43	−1.30	.11
	(−.88)	(−.82)	(−.80)	(.48)
ED	1489.11	1422.35	8.67	−.68
	(1.21)	(1.15)	(1.39)	(−.80)
Sex	17277.56*	27578.43*	129.34*	−3.77
	(2.65)	(2.97)	(2.75)	(−.55)
Race	13388.16**	15674.91*	66.05**	−16.57*
	(1.86)	(2.11)	(1.76)	(−3.02)
BEN	22.87	−32.67	.20	−.01
	(.09)	(−.13)	(−.15)	(−.03)
OCC	−9101.03	−16189.17*	−75.20**	18.96*
	(−1.30)	(−1.96)	(−1.82)	(3.14)
F	4.81*	5.39	4.93	5.62
R^2	.22	.25	.23	.23
No. of observations	196	191	193	193

* Statistically significant at the 5 percent level.
** Statistically significant at the 10 percent level.
( ) t-statistics.

Table 5.6
Impact of Prenotification on the Cost of Displacement,
Interaction between Prenotification
and Local Labor Market Conditions
(Two-Stage Least Squares)

Dependent variable	I Cost of displacement	II Cost of displacement
Constant	−22997.85	−21597.12
	(−.71)	(−.67)
PN*AE	1.66	—
	(.01)	—
PN*IE	—	−12881.59
	—	(−.75)
AE	10.56	12.21
	(.18)	(1.05)
IE	4356.14*	13048.73
	(2.20)	(1.10)
YRLO	−13194.87*	−14141.50*
	(−2.06)	(−2.31)
YRWKD	−1426.01*	−1403.50*
	(−2.71)	(−2.67)
Age	−347.29	−334.68
	(−1.10)	(−1.06)
Ed	1639.34	1503.66
	(1.33)	(1.21)
Sex	20362.53*	21473.31*
	(2.87)	(3.20)
Race	14238.92**	14189.66**
	(1.94)	(1.94)
BEN	2.76	10.95
	(.01)	(.04)
OCC	−11352.34	−12955.78**
	(−1.59)	(−1.77)
F	5.25	5.31
R^2	.24	.25
No. of observations	191	191

* Statistically significant at the 5 percent level.

** Statistically significant at the 10 percent level.

( ) t-statistics.

Table 5.7
Impact of Prenotification and Early Departure
on the Reemployment Success of Displaced Manufacturing Workers
(Two-Stage Least Squares)

Dependent variable	I Cost of displacement	II Change in nominal weekly wage	III Weeks unemployed
Constant	−10960.62	−42.00	52.53**
	(−.28)	(−.21)	(1.70)
PN&QT	27311.08*	123.46*	−.21*
	(2.30)	(1.97)	(−2.30)
AE	2.70	−.01	−.02
	(.20)	(−.14)	(−1.43)
IE	721.53	4.18	−1.81
	(.32)	(.35)	(−1.03)
YRLO	−887.35	−15.37	−15.55*
	(−.12)	(−.39)	(−2.71)
YRWKD	−1414.90*	−7.48*	.49
	(−2.32)	(−2.34)	(1.03)
Age	−170.96	−.89	.18
	(−.45)	(−.44)	(.62)
ED	1379.52	10.74	.71
	(.92)	(1.38)	(.67)
Sex	28856.27*	133.54*	−3.63
	(3.59)	(3.16)	(−.56)
Race	4075.48	2.28	−16.57*
	(.49)	(.05)	(−2.44)
BEN	−364.78	−1.92	−.19
	(−1.21)	(−1.22)	(−.80)
OCC	−3082.68	2.46	21.06*
	(−.34)	(.05)	(3.01)
F	3.58	3.03	4.70
R^2	.27	.24	.29
No. of observations	119	120	139

* Statistically significant at the 5 percent level.
** Statistically significant at the 10 percent level.
( ) t-statistics.

worker left prior to the closing and 0 when a prenotified worker stayed. As expected, workers who are prenotified and early quitters experience smaller displacement losses by $27,311, smaller weekly wage losses by $123, and shorter unemployment periods by .21 weeks (or one day). These preclosing leavers are probably among the most reemployable of the prenotified workers, and advance warning of the closing appears to give many of them an opportunity to leave early. Since 9 percent of displaced workers quit early after prenotification, compared to a quit rate of 1.6 percent over the same period for all manufacturing, it appears that many of the early quits are, in fact, a response to advance notification (U.S. Department of Labor, 1981 and 1982). For workers displaced in tight labor markets, an early departure from a job may also decrease the competition for available jobs and improve workers' reemployment chances.

Summary

We began this chapter by addressing three questions. (1) What impact does local labor market growth have on the readjustment of displaced manufacturing workers? (2) Are there specific demographic groups who experience large financial losses after displacement even when laid off in growing labor markets? (3) What impact does worker prenotification have on the reemployment success of displaced manufacturing workers?

The first two questions can be addressed using the coefficients reported in column I, table 5.4 and several profiles of displaced workers. We select a 50-year-old, white, male, blue-collar worker, displaced in 1979 or 1981, with a 12th grade education, who worked at that firm 25 years prior to layoff, who is receiving the average UI compensation of $90.55 per week and is located in a labor market that lost an annual average of 3,000 jobs in all sectors and 500 industry jobs. This worker would experience an estimated financial loss of $52,000 in the five years immediately following layoff. The same individual in a labor market with an expansion of 3,000 jobs in all sectors and 500 industry jobs would have been only approximately $4,400 better off.

Younger, blue-collar workers also experience financial losses in both growing and declining labor markets. For example, a 25-year-old, male, blue-collar worker, with three years committed to the firm prior to layoff but with the same remaining characteristics as the individual in the declining economy described above, would have lost approximately $16,000 as a result of displacement.

The losses for an older, white-collar male worker within the declining economy and with the same characteristics described in case one above would be approximately $42,000. However, the losses for young, male, white-collar workers are small and may well be compensated for by unemployment insurance and other benefits. For example, the losses for the young, white-collar worker (25 years old, three years committed to the job prior to layoff, with a 12th grade education) in the declining labor market defined above, would have been an estimated $6,200.

The only women who appear to experience large losses are the older, blue- and white-collar workers who committed many years to their employer prior to layoff. These women appear to experience large losses in both growing and declining labor markets. Young blue- and white-collar women with at least a high school diploma appear to make the transition to reemployment with little financial loss.

It should be recognized that the above losses and gains are rough estimates. First, the estimates of displacement costs do not account for wage increases once a worker is reemployed, nor do they account for any wage increases or reductions workers would have experienced had they remained on their prelayoff job. Second, the losses are underestimated because they exclude losses in fringe benefits. Third, the losses may be overestimated because they do not include unemployment benefits or severance payments. The value of UI benefits varies by state and the calculation of likely payments per worker is beyond the scope of this study. However, a rough low estimate of the average benefits paid would range between $2,272 and $3,024.[5] These benefits would cancel the losses experienced by some, but not all, displaced workers.

It should be noted that the model predicts only 22 percent of the variation in the dependent variable. This result is, however, similar to the results from other studies that predict the behavior of individuals.[6] Finally, it should be remembered that the population studied here is comprised of workers displaced in large, relatively diversified local economies. The personal financial losses attributable to displacement may be much larger in small towns and rural labor markets where there are fewer employment opportunities.

In summary, our findings indicate that while labor market conditions influence a displaced worker's reemployment experience, the effects of local labor market conditions are not sufficiently strong to overshadow the impact of years worked prior to layoff and age, occupation, race and sex. As a consequence, most blue-collar workers and mature white-collar workers experience substantial financial losses after displacement, even when displaced in an expanding local economy.

Finally, we did not find evidence that prenotification of layoff reduces financial losses. This result may be due, in part, to the fact that only about 9 percent of the workers in our sample who were prenotified left their jobs prior to the plant's closure. While there is no evidence that prenotification alone improves the displaced workers' income stream, this should not be interpeted as an argument against advance notice legislation. Prenotifiation gives workers a chance to adjust expenditures in the face of upcoming losses in income, communities a chance to establish job training and counseling programs to assist displaced workers, and communities and workers a chance to seek alternatives to the plant closing.

NOTES

1. Jacobson and Thomason (1979) and Holen, Jehn, and Trost (1981) tracked the difference between the trajectory of prelayoff wages and reemployment wages and found that the gap did in fact narrow over time. In contrast, Ruhm (1987) examined the persistence of wage losses for involuntarily laid-off workers and found 42 percent did not experience any narrowing of the loss after five years.

2. In all calculations where the present value of the unemployment experience is used, the sample universe excludes the self-employed and workers who have dropped out of the labor force. The self-employed are excluded because wages are not collected for this group. Workers who have dropped out of the labor force are excluded because we cannot value their current wage at zero, as we do with the involuntarily unemployed. These are individuals who may put a value on leisure or other nonpaid productive work that exceeds the value of paid work in the labor force. This group is most likely to include workers near retirement age; women, especially those with young

children; and young workers who are returning to school. This group also includes discouraged workers, who probably experience the largest financial losses from displacement. The costs of displacement for those on the work-no work margin are probably low, whereas the costs of displacement for the discouraged worker are high. To the extent these two groups offset each other, there should be no bias in CD as a result of the exclusion of workers who have dropped out of the labor force.

3. Testimony by James Savoy, President Local Lodge 2689 International Association of Machinists, before the U.S. Congress (1983).

4. The BLS does not report a quit rate for non-notified workers.

5. The average manufacturing worker was unemployed longer than 26 weeks, the maximum period for receiving UI benefits under the state UI system. The average weekly benefits paid in all industries was $89.68 in 1979 and $119.34 in 1982 (U.S. Department of Labor 1983, p. 208). The above value is equal to the present value of 26 weeks of benefits at these levels. This is estimated to be a lower bound on UI benefits because many displaced manufacturing workers are eligible for extended benefits and because manufacturing wages are higher than the all-industry average. Consequently, the average weekly benefits paid to manufacturing workers is probably higher than the above estimate.

6. For example, in a regression equation with similar independent variables plus home ownership, Bendick and Devine (1981) reported an R^2 of .02. Holen, Jehn, and Trost (1981) found demographic characteristics of displaced workers explained only 20 percent of the variation in wage losses one year after displacement. Clearly, variables other than the ones included here influence the personal financial costs associated with displacement. Possible missing variables are personal attributes of the workers.

Appendix 5.1
Data Sources for the Labor Market Variables

Total Employment Growth, March 1979 to March 1982 data: U.S. Department of Labor, Bureau of Labor Statistics, *Employment and Earnings*, June 1980 and May 1983.

Industry Employment at 2-Digit SIC Code Level, 1977 to 1981 data: U.S. Department of Labor, Bureau of Labor Statistics, *Supplement to Employment and Earnings, States and Areas, Data for 1977-81*, Bulletin 1370-16, September 1982. 1980-1983 data: U.S. Department of Labor, Bureau of Labor Statistics, *Supplement to Employment, Hours and Earnings, States and Areas, Data for 1980-84*, Bulletin 1370-19, September 1985.

Manufacturing Employment, 1979 to 1982 data: U.S. Department of Labor, Bureau of Labor Statistics, *Employment and Earnings*, table B-8, March 1980 and table B-8, February 1984.

Mean Unemployment Rate, 1979 to 1982: U.S. Department of Labor, Bureau of Labor Statistics, *Employment and Earnings*, June 1980 and May 1983.

Appendix 5.2
Pearson Correlation Coefficients
Among Selected Variables

	ALLEMP	MANU	UE	IND	UECH	AGE	PLWAG
ALLEMP	1.0	.12	-.59	.03	-.19	—	—
MANU	—	1.0	-.64	.69	-.40	—	—
UE	—	—	1.0	-.46	.35	—	—
IND	—	—	—	1.0	-.26	—	—
AGE	—	—	—	—	—	1.0	.14
YRSWKD	—	—	—	—	—	.56	.23
ED	—	—	—	—	—	—	.29
SEX	—	—	—	—	—	—	-.41
BEN	-.25	-.55	.29	-.36	.28	—	—
RACE	—	—	—	—	—	—	.15

SOURCE: U.S. Bureau of Labor Statistics, Current Population Survey of Displaced Workers, January 1984.

REFERENCES

Aiken, Michael, Louis Ferman, and Harold Sheppard (1968) *Economic Failure, Alienation, and Extremism.* Ann Arbor, Michigan: University of Michigan Press.

Bendick, M., Jr. and J.R. Devine (1981) *Workers Dislocated by Economic Change: Do They Need Federal Employment and Training Assistance?* seventh annual report, The Federal Interest in Employment and Training, National Commission for Employment Policy. Washington, D.C., October, pp. 177-226.

Classen, Kathleen P. (1977) "The Effect of Unemployment Insurance on the Duration of Unemployment and Subsequent Earnings," *Industrial and Labor Relations Review* 30, 4, pp. 438-444.

Conant, Eaton H. (1965) "Report and Appraisal: The Armour Fund's Sioux City Project," *Monthly Labor Review* 88, 11 (November), pp. 1297-1301.

Dorsey, John (1967) "The Mack Case: A Study in Unemployment," in Otto Eckstein (ed.), *Studies in the Economics of Income Maintenance.* Washington, D.C.: The Brookings Institution, pp. 175-248.

Ehrenberg, Ronald (1985) "Worker's Rights: Rethinking Protective Labor Legislation," unpublished paper, Cornell University, Ithaca, New York.

Feldstein, Martin (1974) "Unemployment Compensation: Adverse Incentives and Distributional Anomalies," *National Tax Journal* 27, 2, pp. 231-244.

Flaim, Paul O. and Ellen Sehgal (1985) *Displaced Workers of 1979-83: How Well Have They Fared?* Bulletin 2240, U.S. Bureau of Labor Statistics. Washington, D.C.: Government Printing Office, July.

Folbre, Nancy, Julia Leighton, and Melissa Roderick (1984) "Plant Closings and Their Regulation in Maine, 1971-1982," *Industrial and Labor Relations Review* 37, 2, pp. 195-196.

Foltman, Felician (1968) *White and Blue-Collars in a Mill Shutdown.* Ithaca, New York: ILR Press.

Gainer, William (1986) "U.S. Business Closures and Permanent Layoffs During 1983 and 1984," paper presented at the Office of Technology Assessment–General Accounting Office workshop on plant closings, April 30–May 1, 1986.

Gordus, Jeanne Prial, Paul Jarley, and Louis Ferman (1981) *Plant Closings and Economic Dislocation.* Kalamazoo, Michigan: W.E. Upjohn Institute for Employment Research.

Hammerman, Herbert (1964) "Five Case Studies of Displaced Workers," *Monthly Labor Review* 87, 6 (June), pp. 663-670.

Holen, Arlene, Christopher Jehn, and Robert P. Trost (1981) *Earnings Losses of Workers Displaced by Plant Closings.* Alexandria, Virginia: Public Research Institute, Center for Naval Analysis, CRC 423, December.

Horvath, Francis (1982) "Forgotten Unemployment: Recall Bias in Retrospective Data," *Monthly Labor Review* 105, 3 (March), pp. 40-44.

Jacobson, Louis (1977) *Earnings Losses of Worker Displacement When Employment Declines in the Steel Industry,* Ph.D. dissertation, Northwestern University.

Jacobson, Louis and Janet Thomason (1979) *Earnings Loss Due to Displacement.* Alexandria, Virginia: Public Research Institute, Center for Naval Analysis, CRC 385, August.

Lipsky, David B. (1970) "Interplant Transfer and Terminated Workers: A Case Study," *Industrial and Labor Relations Review* 23, 2, pp. 191-206.

Maddala, G.S. (1977) *Econometrics.* New York: McGraw Hill.

McKenzie, Richard (1982) "Frustrating Business Mobility," in Richard McKenzie (ed.) *Plant Closings: Public or Private Choices?* Washington, D.C.: Cato Institute, pp. 7-18.

O'Connell, Mary (1985) "Playskool Settlement: So Far Not So Good," *The Neighborhood Works* 8 (September), p. 1.

Palen, John and Frank J. Fahey (1968) "Unemployment and Reemployment Success: An Analysis of the Studebaker Shutdown," *Industrial and Labor Relations Review* 21, 2 (January), pp. 234-250.

Podgursky, Michael and Paul Swaim (1987) "Job Displacement and Earnings Loss: Evidence from the Displaced Worker Survey," *Industrial and Labor Relations Review* 41, 1, pp.17-29.

Ruhm, Christopher (1987) "The Economic Consequence of Labor Mobility," *Industrial and Labor Relations Review* 41, 1, pp. 30-42.

Shultz, George P. (1964) "The Fort Worth Project of the Armour Automation Committee," *Monthly Labor Review* 87, 1 (January), pp. 53-57.

Stern, James L. (1969) "Evolution of Private Manpower Planning in Armour's Plant Closing," *Monthly Labor Review* 92, 12 (December), pp. 21-28.

Stern, James L. (1971) "Consequences of Plant Closure," *Journal of Human Resources* 7, 1 (December), pp. 3-25.

Tresch, Richard W. (1981) *Public Finance, A Normative Theory.* Georgetown, Ontario: Irwin, Dorsey.

Ullman, Joseph C. (1969) "Helping Workers Locate Jobs Following a Plant Shutdown," *Monthly Labor Review* 92, 4 (April), pp. 35-40.

United States Department of Commerce, Bureau of the Census (1985) *Statistical Abstract.* Washington, D.C.: Government Printing Office.

U.S. Congress (1983) Testimony before the Subcommittee on Labor Management Relations of the Committee on Education and Labor, 98th Congress, First Session, HR 2847. Washington, D.C.: Government Printing Office.

United States Department of Labor, Bureau of Labor Statistics (1981 and 1982) *Employment and Earnings.* Washington, D.C.: Government Printing Office, December 1981, Table D-1 and March 1982, Table D-1.

United States Department of Labor, Unemployment Insurance Service (1983) *Employment and Training Handbook #394,* Unemployment Insurance Financial Data. Washington, D.C.: Government Printing Office, October 14.

Welch, Finis (1977) "What Have We Learned from Empirical Studies of Unemployment Insurance?" *Industrial and Labor Relations Review* 30, 4, pp. 451-461.

Wendling, Wayne R. (1984) *The Plant Closure Policy Dilemma.* Kalamazoo, Michigan: W.E. Upjohn Institute for Employment Research.

6

Summary, Policy Implications and Directions for Further Research

This study of regional plant closure patterns in three manufacturing industries and of the costs of displacement in all manufacturing was designed to identify (1) where plant closures are most likely to occur, (2) the causes of plant closures, and (3) the conditions under which plant closure and worker displacement lead to large financial losses for the displaced worker. Chapter 6 summarizes the major findings and their immediate policy implications and reviews several directions for further research.

Data on the three manufacturing industries were extracted from the Dun and Bradstreet Dun's Market Identifiers file for the years 1973, 1975, 1979, and 1982. The three industries include metalworking machinery (SIC 354), electronic components (SIC 367), and motor vehicles (SIC 371). In total, there were approximately 54,000 establishments in the data set, representing a 100 percent sample of total employment in the three industries as measured by the *Census of Manufactures*. One strength of these data is that employment growth can be disaggregated into components of employment change, including employment growth due to plant start-ups, closings, net expansions or contractions, and plant migrations.

Plant Closures and Regional Economic Growth

Three sets of findings are drawn from the analysis of the D&B data. First, there is little regional variation in rates of plant closure. The range (difference between the highest and lowest rate) is 2.3 percentage points in metalworking machinery, 2.9 percentage points in electronic components, and 2.8 percentage points in motor vehicles. Moreover, there is no correlation between cross-regional rates of industry growth and regional rates of plant closures.

In contrast, the regional variation in rates of employment gains due to start-ups can be large and is highly correlated with cross-regional differences in industry growth rates. In metalworking machinery, growth rates vary from –1.9 percent in the Mid-Atlantic to 9.5 percent in the Mountain states, while start-ups range between 1.6 percent to 10.3 percent. In electronic components, growth rates vary from –2.6 percent in the Mid-Atlantic region to 17.7 percent in the East South Central, and start-ups in these same regions vary from 2.4 percent to 17.5 percent. Only in motor vehicles is the correlation between employment change and start-ups weak, but this is primarily because there is very little cross-regional variation in either of these statistics. Growth varies from –2.9 percent in the Pacific to 1.3 percent in the East South Central, while start-ups vary from .5 to .3 percent in these regions.

It is clear that regions with slow growth have low start-up rates and limited expansions. They do not have particularly high rates of plant closures. Conversely, fast growth is due primarily to high start-up rates, rather than to low rates of job loss due to closures.

These results are consistent with the findings of Birch (1979) of MIT and Armington and Odle (1982) of Brookings, who reported similar findings for manufacturing as a whole and other major economic sectors.

Second, we further tested the relationship between regional economic growth and rates of plant closures in a model estimating the probability of a closure during the period 1975 to 1982. The independent variables included establishment size, age, status as a branch, subsidiary, headquarters, or independent operation, and regional and interregional location. Few of the locational dummies were statistically significant, and when they were significant they indicated the probability of closure was higher for establishments located in fast-growing southern regions than for those in the slow-growing or declining northern regions.

Third, we tested the view that relatively high production costs are responsible for many plant closings by examining the effect on plant closings of several local economic factors—the level and rate of change of wages and utility rates, the level of taxes, the extent of unionization, the degree of import competition, and changes in product demand. We also examined the effect of plant size and status as a branch, subsidiary, headquarters, or independent. Because the data on local economic conditions are only available by Standard Metropolitan Statistical Areas (SMSAs), our sample was limited to plants in SMSAs.

We found status as a branch plant or subsidiary influences closures in all three industries. A branch plant is 32 percent, 25 percent, and 16 percent more likely to close than is an independent in the metalworking, electronic components, and motor vehicle industries, respectively. Comparable statistics for subsidiaries versus independents in the same three industries are 17 percent, 8 percent, and 11 percent. On the other hand, independents and headquarters are equally likely to close.

When the economic variables are included in the equation, establishment size has a significant effect on closures in metalworking machinery and motor vehicles, with larger plants closing at lower rates. But in 19 of the 21 tests, economic variables were shown to have no significant effect on plant closures. In two cases, there was a significant effect but in the "wrong" direction. In the metalworking industry, low wages and increases in local product demand were associated with an increased probability of closure.

These conclusions contradict the results of some survey-based studies that interviewed managers responsible for plant closures. For example, in Gainer's (1986) survey of 355 firms that closed a plant, 57 percent cited the high cost of labor as a major factor in their decision. Schmenner (1982) discovered that 21 percent of managers attributed their plant shutdowns to high labor costs and 10 percent cited labor unions.

Two explanations for the inconsistency between our statistical results and the findings of surveys are possible. One is that managers attempt to exert downward pressure on wages in general, and in their other continuing plants, by overstating the role that wage rates played in a closure. The closure of a plant may be used as an opportunity to intimidate labor and encourage workers to rein in their future wage demands. Another possible explanation is that wages are a large portion of a firm's costs and are especially noticeable when the plant falls on hard economic times. High labor costs, rather than more subtle reasons for closure, such as the company's failure or inability to adopt labor-saving technologies, find new markets, improve worker productivity, improve the quality of the product in the face of competition, etc., may be blamed for the closure.

The impact of unionization as a cause of plant closures also appears to be exaggerated. Scholars from all political perspectives, as well as business spokespersons, frequently attribute plant closures to justified or unjustified demands by unions (see, for example, Bluestone and Harrison 1982). High wages, relatively generous fringe benefits, costly strikes, and health and safety restrictions can all accompany a unionized labor force and supposedly reduce a plant's competitiveness. While there is evidence that the small number of metalworking machinery plants which do relocate tend to move out of unionized regions, there is no indication that plants situated in a unionized labor market are more likely to close than plants located in a nonunion area.

There are two theories for the relatively high closure rates of branches and subsidiaries. One theory argues that managers of multidivisional firms and subsidiaries expect branches to yield higher rates of return than is the case for single-plant entrepreneurs. This hypothesis is plausible because multiplant firms have good information about which branches and divisions are most profitable. Consequently, they can and do allocate cash flow to the highest yield activities and phase out, divest themselves of, and close the least profitable operations. In addition, multiplant firms also have the expertise and staff to search outside the current product line and locations for more profitable sites and ventures, again phasing out less profitable activities.

In contrast, entrepreneurs operating single-plant firms have fewer investment alternatives. They will be more likely, therefore, to keep a plant operating when its profits fall below the rate at which a multiplant firm would close down the operation. Nonpecuniary factors may also play an important role. The small scale entrepreneur may enjoy running the company and is likely to live in, derive status from, and have a commitment to the community where the business is located.

The second view also argues that plants are closed when profits are inadequate, but in this model multiplant firms are responsible for the low profitability of their branches and subsidiaries. One explanation of why profits are low in the long run is that corporations expend resources on short-run paper profits, rather than on making the investments needed to develop new products, find new markets, and improve production technologies. A second is that corporations frequently acquire activities in which they have little operating experience. As a result, a once profitable acquisition may be poorly managed, become unprofitable, and ultimately be closed.

Policy Implications for Local Economic Development

Regardless of which explanation best fits a particular branch or subsidiary closing, the above results suggest that local takeovers of "unprofitable" branches and subsidiaries can be an option for keeping plants operating and jobs in the community. This is a viable strategy because (1) closures do not appear to be the inevitable consequence of local economic conditions, and (2) local new owners may either be willing to accept lower rates of return in exchange for community and job stability or be able to improve upon the previous management to establish a satisfactory rate of return.

While there is little literature on local private takeovers of closing plants, there is a burgeoning interest in employee takeovers. Evidence on the profitability of existing employee-owned firms is encouraging. A number of studies have compared profitability in employee-owned firms and producer cooperatives with profits in privately-owned firms in the same industry. Conte (1982) surveyed a number of these studies, identifying their shortcomings. A number of difficulties in comparing profitability in privately- and employee-owned firms arise. For example, profits in an employee-owned firm may be paid in wages and thus not show up on the balance sheet as profits. Another difficulty with such comparisons is that many successful employee-owned firms are eventually taken over by private companies. As an employee-owned firm's profits and stock values increase, currently employed workers often cannot afford to purchase the stock of retiring employees. Rather than sell stock at below-market prices, employees sell to private interests. In other cases, successful employee-owned firms have gone public to acquire expansion capital.

Taking these difficulties into account, Conte concludes that the average employee-owned firm is as profitable, if not more so, than the average privately-owned firm in the same industry. Although experience with employee-owned firms is still limited, several case studies indicate that both improved efficiency and wage concessions, combined with new management, are responsible for the successes (Bradley and Gelb 1985; Whyte 1986; *Washington Post* 1985).

Rosen, Klein, and Young (1986, p. 27) of the National Center for Employee Ownership estimate there have been 65 buyouts since 1971. About 90 percent are still in business. Out of the 65 buyouts, 2 are coming out of Chapter 11 bankruptcy proceedings, 4 have been profitably sold to conventional businesses, and 5 have closed.

There are a number of policies the federal government can adopt to facilitate local takeovers of closing establishments. One barrier to local ownership is a lack of capital. Local banks are the major source of capital for small business, and they are often reluctant to risk financing a business that has a poor track record and a reputation for failure. In the case of employee takeovers, many banks have also been reluctant to make loans to an ownership form with which they are unfamiliar. Moreover, if the closing operation is large, a local bank may not have the resources to make the loan and still keep a diversified portfolio. Federal loan guarantees of up to 80 percent of the value of a loan would reduce the risk to banks and encourage banks to make the loans, leaving the banks with enough at stake to carefully evaluate the business' viability. In a number of instances, federal loan guarantees have already been critical factors in employee buyouts.

A second federal initiative that would facilitate local takeovers is mandatory prenotification of a closing. In order for workers and communities to plan a buyout, they must have advance warning that the business is closing. The chances for a successful takeover are much greater when new owners can purchase an operating business. Once managers have departed and ties with suppliers and markets have been severed, returning a company to profitability is difficult. In order to purchase operations with markets and employees intact, local buyers must have time to conduct feasibility studies and negotiate financing. While the length of time necessary for planning a local takeover varies by the size of plant, case studies of worker buyouts of larger plants suggest that at least 180 days are needed to negotiate and assemble financing (Stern, Wood, and Hammer 1979; Whyte 1986; Logue, Quilligan, and Weissmann 1986).

Federal initiative in this area is critical. States are reluctant to take the lead in requiring prenotification for fear of discouraging new investment. With federal prenotification legislation as an umbrella, existing state programs could improve their success rates.

The above findings also suggest state and local governments should look critically at claims by the private sector that high taxes are driving plants out of business or that tax concessions will put a plant operating in the red into the black. We found no evidence that plant closures were more frequent events in high-tax than low-tax areas. In fact, tax cuts may adversely affect local development if the quality of the local infrastructure and schools is diminished as a consequence. Similarly, utility companies should avoid pricing policies that shift costs from manufacturers to consumers as a job retention strategy, and employees should be skeptical of wage concessions as a means of saving jobs. Where firms demand such concessions, it is likely that companies are taking advantage of a strong bargaining position due to the need for local jobs to wrest tax, utility rate and wage concessions from government, utility companies and employees. While surely wage, utility rate, and tax reductions will improve the profit performance of a particular plant, there is no evidence that such policies alone will reduce local rates of plant closures.

This conclusion on wage concessions is supported by other empirical evidence. In a study of the tire industry, Cappelli (1985) found that "virtually every one of the plants in which concessions were negotiated in recent years eventually closed." The same author also argues that firms are threatening shutdowns and engaging in concession bargaining when there is no danger of closing. For example, a poll of 619 large, unionized employers found that 11 percent of the firms surveyed were taking advantage of the fear of plant closures to demand employee concessions, even though the company did not need them to survive. In 1984, the *Wall Street Journal* also reported that a number of profitable firms have used the threat of outsourcing and shutdowns to force concessions from unions (Cappelli 1985, p. 101).

Two caveats should be added to these conclusions. First, our results and policy conclusions only hold for the range of wages, taxes, utility costs, etc., currently found across SMSAs. If a locality, for example, were to raise tax rates substantially above the range of values tested in this study, closure rates could rise. Second, although we found no evidence that high wages, utility costs, and rates of unionization encouraged plant closures, these conditions have been found to discourage new investments in the form of plant start-ups and expansion. Consequently, the above policy conclusions should not be interpreted as a suggestion that wages and other costs can be raised without adverse impact on local economic growth.

Cross-Industry Variations in Plant Closures

Whereas plant closure rates do not vary much across the regions, neither do they appear to vary substantially across industries. To the extent they do vary at all, they are positively associated with industry employment growth. The 1973-79 rate of job loss due to closures was −3.0 in metalworking machinery, while the overall rate of national growth was .2, according to the D&B data. In electronic components, which grew at an annual average rate of 1.7 percent, the rates of job loss due to plant closures was −3.9 percent. The comparable figures for motor vehicles were −2.3 for closure and −2.4 for growth.

The result is similar for the 4-digit industries within these three 3-digit industries. There was little variation in plant closure rates across industries, after plant size and status are held constant.

We also explored the impact of imports on job loss due to plant closures and found no evidence that plant closure rates are higher in industries facing strong import competition. While clearly these conclusions are limited to metalworking machinery, electronic components, and motor vehicles, these industries vary widely in their rates of growth and vulnerability to imports. For example, within motor vehicles (SIC 371), imports in motor vehicle bodies (SIC 3711) grew by 8 percent per year, while truck body imports (SIC 3713) did not grow at all. Yet plants in both industries were equally likely to close, after plant status and location were held constant. Similar statistics characterize metalworking machinery and electronic components. Our findings are consistent with unpublished data from the Brookings Institution's Small Business Microdata Project. Brookings used the total D&B file and found that cross-industry variations in manufacturing start-ups and expansions, rather than cross-industry variations in closures, were responsible for variations in industry rates of growth at the 2-digit level of industrial detail.

Industrial Policy as a Strategy to Assist Displaced Workers

The findings raise several questions about proposals to assist displaced workers in declining industries. Import restrictions and government subsidies to declining industries are two proposed strategies in the ongoing debate on industrial policy and plant closing legislation.[1] The intention of industrial protection is to slow exits from the industry and give firms time to diversify product lines or update their capital stock. The responsiveness of start-ups to market growth suggests that government policies to protect an industry and increase market demand may encourage new plant openings in the industry rather than reduce the rate at which existing establishments are phased out. This will not aid displaced workers unless the new jobs are created in the same labor market where the plant closings occur. Given the current shift of new investment to the Sun Belt, this outcome is unlikely. The textile industry is an example where years of import protection helped textile companies, but did not preserve jobs for New England's textile workers.

It should be noted, however, that these conclusions and policy implications are limited to workers displaced by plant closures and do not extend to permanent layoffs in continuing plants. Permanent layoffs in on-going establishments may be greater in import-affected and declining industries, a question not explored in this study.

Financial Costs of Worker Displacement

As stated above, plant closures occur at almost even rates across the nation's regions. The issue of whether worker displacement is or is not a national issue depends on (1) whether all displaced workers, even those laid off in growing markets, experience large financial losses, or (2) whether the losses in wages and long unemployment spells are problems specific to dislocated workers in declining labor markets.

We used the Bureau of Labor Statistics' Current Population Survey— Supplement on Displaced Workers to examine financial costs of worker displacement. The survey identified workers who had been involuntarily displaced from their jobs between January 1979 and January 1984. Analysis of these data shows that workers displaced in local markets with a growing demand for labor experience shorter unemployment spells and smaller wage losses than workers laid off in weak labor markets. However, most displaced workers, even those displaced in strong labor markets, experience some financial loss in the five years after layoff. For example, we found the average metalworking machinery worker in a declining labor market experienced $17,182 in financial losses in the five years after displacement. The average metalworking machinery worker displaced in a growing labor market experienced an average of $1,173 in financial losses. The equivalent figures for workers displaced from transportation manufacturing are $45,942 in declining and $6,130 in growing labor markets, and for all manufacturing workers the estimted losses are $31,298 for workers displaced in declining labor markets and $19,759 for workers displaced in growing areas.

As anticipated, older and less-educated workers are hardest hit by the permanent loss of their jobs, a finding consistent with numerous other studies. Contrary to others, however, we found that women experience smaller financial losses after dislocation than men. The losses to a displaced woman, in the five years after layoff, were an estimated $19,000 lower than losses experienced by a man of the same age, educational attainment, race, and labor market. The difference in the male and female postlayoff experience is explained by the tendency for women to come closer to recouping their prelayoff wage, rather than by shorter periods of unemployment for women than men. The smaller losses for women may be explained either by women's lower prelayoff wage, which is easier than men's to duplicate, or the fact that women are more likely than men to hold clerical jobs in the manufacturing sector, and consequently to have skills valued in the growing service sector.

Policies to Assist Displaced Workers

Even with adequate legal provisions for employee prenotification and assistance for employee ownership, many establishments cannot be made profitable and jobs cannot be saved. The results presented above further emphasize the findings of many other researchers, which argue for additional adjustment assistance to older workers, and for increased investment in education to ease labor adjustments to economic change.

Assistance for older displaced workers may include phased-in early social security payments, legislated early retirement benefits for older workers, or extended unemployment insurance benefits. Because job search skills for this group of workers are rusty at best, resume writing and job search workshops have been proposed, along with support groups or "job clubs" which offer displaced workers emotional support and companionship during the adjustment to a new job or early retirement.

Efforts to promote formal education should also reduce the costs of displacement. Programs to encourage young people to remain in school before entering the workforce, to attend continuing education programs, or to return to school after displacement, should also reduce the personal and societal costs of displacement. The effort here should be on reading, writing and analytical skills to complement the $30 to $50 million annual expenditures made by private industries for education (Bendick 1983). Data from the displaced workers survey show that 10 percent of displaced workers have attended elementary school only. Another 15 percent have attended high school but have not graduated. Efforts should be made to facilitate additional education for these workers, by paying unemployment insurance benefits to them while in school, for example, and permitting tuition tax credits for those training for jobs outside their current profession.

A final finding from these data is that workers who have advance notice of a closure do not appear to experience smaller financial losses

than workers who are caught by surprise. One explanation is that even
with advance notice, few workers leave prior to the plant closure. Only 9 percent of prenotified workers left before their layoff. As expected,
however, the prenotified early leavers experience the smallest financial losses. While prenotification may not improve incomes, it does give
workers a chance to adjust expenditures in anticipation of a period of
reduced income, and, as argued above, gives workers and communities
an opportunity to explore the feasibility of a local plant buyout.

Directions for Further Research

The study of plant closure rates across regions and SMSAs has cast
doubt on a number of assumptions about the causes of closures, but
has provided little explanation of why plants close. Our ability to predict
where and why a plant will close is still limited. Future research might
focus on in-depth case studies of continuing and closing plants and
develop alternate and testable hypotheses.

A second area for further study is to look at plant closure patterns,
rates of job displacement, and the costs of layoff across industries rather
than regions. Contrary to accepted wisdom, we found that plant closure
rates in 4-digit industries were unaffected by the degree of import competition and industry growth. This finding is limited, however, by the
inclusion of only three 3-digit SIC code industries. Research which includes data on all industries and tests the hypotheses with other data
sets can help establish the validity and generalizability of this result.

Analyses using all industries would also test our speculation that it
isn't until an industry's start-ups and expansions are very low, that losses
due to plant closures begin to increase and lead to further industrial
decline. Moreover, we did not consider the probability of layoff by industry. While this study, combined with the results of the Brookings'
studies, suggests that layoffs are as likely in declining as growing industries, this question cannot be carefully tested with the D&B data.
The D&B data only allow us to determine whether there was a job loss
and not whether that job loss is temporary or permanent.

An across-industry study of the costs of displacement to workers is a third area for study with potential policy implications. For example, the results of such a study would be valuable in evaluating the Trade Adjustment Assistance (TAA) program. TAA is designed to aid workers displaced by increased foreign competition; the benefits are cash payments made available after unemployment insurance expires. The Trade Expansion Act of 1962, later modified in 1974 and 1981, determined that workers were eligible for benefits where imports made an important contribution to their separation from their jobs. The purpose of TAA is not only to buy political support for trade liberalization, but to assist workers believed to encounter the greatest reemployment barriers. An explicit assumption in the TAA program is that workers in declining industries experience the largest losses from layoff, presumably because they are the most likely to be laid off or, if laid off, the least likely to find a comparable job.

Preliminary findings from this study indicate there is no simple relationship between industrial decline and layoffs. The question remains as to whether workers from declining industries experience larger financial losses because of limited reemployment opportunities. One hypothesis is that TAA is inequitable. Under TAA, a worker displaced from an industry exhibiting national growth but local decline is not eligible for assistance while a worker displaced from an industry exhibiting national decline but local growth is eligible. The former may face greater displacement losses and reemployment barriers than the latter, and yet is not eligible for TAA.

Evidence arguing against industry-specific programs is provided by Bendick and Devine (1981), who compared the length of the unemployment spell of workers displaced from declining industries with that of workers displaced from growing industries. They found that 12.1 percent of workers who had previously worked in growing industries were still unemployed after 26 weeks, while only 10.7 percent of workers displaced from declining industries were still unemployed after 26 weeks. This pattern was reinforced with regression analysis, with the characteristics of workers held constant. Working in a declining industry had no statistically significant effect on the expected duration of unemployment. One explanation for this result may be that the length of the unemployment spell is more sensitive to local industry growth than national industry growth. This is an area for further work.

A fourth area for future research is in the study of dislocated workers who migrate. One shortcoming of the data analyzed in chapter 5 is that we could not determine where a dislocated worker who had moved lived prior to the relocation. Consequently, it was impossible to determine whether movers experienced smaller financial losses than stayers. Another related task is to identify the demographic and social characteristics of nonmovers in depressed labor markets. An answer to these questions will be helpful in determining whether financial constraints are a barrier to the migration of displaced workers and whether relocation assistance is likely to ease the adjustment for displaced workers.

A fifth and final direction for future research is to study the reemployment experience of workers in nonmetropolitan labor markets. Because of the lack of information on the location of dislocated workers outside of SMSAs, these nonmetropolitan displaced workers had to be eliminated from this study. Data on the reemployment experience of workers displaced in what are probably less diversified economies will broaden our understanding of worker adjustments to job loss.

NOTE

1. See, for example, U.S. Congress (1983) the testimony by F.M. Lunni, Jr., Assistant Vice-President, Industrial Relations, National Association of Manufacturers, pp. 275-278.

REFERENCES

Armington, Catherine and Marjorie Odle (1982) "Sources of Recent Employment Growth, 1978-1980," unpublished paper prepared for the second annual Small Business Research Conference, Bentley College, Waltham, Massachusetts, March 11-12, 1982.

Beauregard, Robert A., Carl E. Van Horn, and David S. Ford (1983) "Governmental Assistance to Displaced Workers: An Historical Perspective," *Journal of Health and Human Resources* 6, 2 (Fall), pp. 166-184.

Bendick, M., Jr. (1983) "Government's Role in the Job Transition of America's Dislocated Workers" A statement before the Committee on Science and Technology and the Committee on the Budget, U.S. House of Representatives, Hearings on Technology and Employment, Washington, D.C., June 9.

Bendick, Mr., Jr. and J.R. Devine (1981) *Workers Dislocated by Economic Change: Do They Need Federal Employment and Training Assistance?* Seventh annual report, The Federal Interest in Employment and Training, National Commission for Employment Policy, Washington, D.C., October, pp. 175-226.

Birch, David (1979) *The Job Generation Process.* Cambridge, Massachusetts: MIT Program on Neighborhood and Regional Change.

Bluestone, Barry and Bennett Harrison (1982) *The Deindustrialization of America.* New York: Basic Books.

Bradley, Keith and Alan Gelb (1985) "Employee Buyouts of Troubled Companies," *Harvard Business Review* (September/October), pp. 121-130.

Cappelli, Peter (1985) "Plant Level Concession Bargaining," *Industrial and Labor Relations Review* 39, 1, pp. 90-104.

Conte, Michael (1982) "Participation and Performance in U.S. Labor Managed Firms," in Derek C. Jones and Jan Svejnar (eds.), *Participatory and Self Managed Firms.* Lexington, Massachusetts: Lexington Books, pp. 213-237.

Gainer, William J. (1986) "U.S. Business Closures and Permanent Layoffs During 1983 and 1984," paper presented at the Office of Technology Assessment-General Accounting Office workshop on plant closings, April 30–May 1.

Logue, John, James B. Quilligan, and Barbara J. Weissmann (1986) *Buyout: Employee Ownership as an Alternative to Plant Shutdowns: The Ohio Experience.* Kent, Ohio: Kent Popular Press.

Neuman, George E. (1981) "Adjustment Assistance for Displaced Workers,"
in Robert Baldwin and J. David Richardson (eds.), *International Trade
and Finance*. Boston, Massachusetts: Little, Brown, pp. 157-180.

Rosen, Corey M., Katherine J. Klein, and Karen M. Young (1986) *Employee
Ownership in America: The Equity Solution*. Lexington, Massachusetts:
Lexington Books.

Schmenner, Roger (1982) *Making Business Location Decisions*. Englewood
Cliffs, New Jersey: Prentice Hall.

Stern, Robert N., K. Haydn Wood, and Tove Helland Hammer (1979) *Employee Ownership in Plant Shutdowns*. Kalamazoo, Michigan: W.E. Upjohn Institute for Employment Research.

United States Congress (1983) Testimony before the Subcommittee on Labor
Management Relations of the Committee on Education and Labor, 98th
Congress, First Session, HR 2847. Washington, D.C.: Government
Printing Office.

Washington Post (1985) February 2, p. D1.

Whyte, William Foote (1978) "In Support of Voluntary Employee Ownership," *Society* (September/October), pp. 73-82.

Whyte, William Foote (1986) "O & O Breakthrough: The Philadelphia
Story," *Society* (March/April).

INDEX

Age of firm: and plant closings, 53, 90, 103

Age of worker: as criterion for reemployment, 12-21, 119-21; and financial losses, 160; *See also* Reemployment

Aiken, Michael, 122

Armington, Catherine, 15, 42, 52, 152

Armour and Company plant closing, 119

Barkley, David L., 52

Belitsky, Harvey A., 90, 91

Bendick, Marc, Jr., 120, 145 n6, 161, 163

Birch, David, 15, 19, 21, 26, 42, 52, 152

Bluestone, Barry, 3, 4, 51, 52, 71, 72, 73, 154

Bradley, Keith, 156

Branch plants: employment in, 52; and plant closings, 10, 49-53, 66, 153-54; *See also* Multiplant firms; Single-plant firms; Subsidiary plants

Bureau of Labor Statistics survey of displaced workers. *See* Current Population Survey

C and R Associates, 51

Capelli, Peter, 157, 158

Carlton, Dennis, 31, 67-68, 106

City Reference File (CRF), 14-15

Classen, Kathleen P., 123

Clinton Colonial Press plant closing, 51

Committee for Economic Development study on industrial competitiveness, 76-77

Conant, Eaton H., 120

Conte, Michael, 155, 156

Conte, Silvio O. (Congressman), 7

Critchlow, Robert V., 91

Current Population Survey (CPS),

Supplement on Displaced Workers, 10, 23-24, 115-19, 160

Cyert, Richard M., 50

Daniels, Belden, 49

Data for analyses: evaluation of, 13-28

Declining industries: and industsrial policy, 159

Demand for product: growth in, 90-92; hypothetical effect of, in plant closings, 77; as variable in probability model, 87-90

Devine, James N., 120, 145 n6, 163

Dicken, Peter, 50

Displaced workers: benefits for, in other countries, 8; concentration of, by region, 47, 115; in declining industries, 11, 159; definition of, 116-17; educational levels of, 161; and estimated financial losses, 123-29, 130-32, 137-44, 160; in Kansas City, 122; in labor market, 128; location shift of, 130; in manufacturing, 127-28; as national issue, 11, 67, 115; and prenotification, 137-44, 161-62; problems experienced by, 2; proposed assistance for, 161; and public policy, 5-8, 11, 151, 156-58, 159; and reemployment, 130, 133; and regional change, 11; *See also* Financial losses; Job loss

Dorsey, John, 121, 122

Dresser Industries plant closing, 135

Duracell plant closing, 67

Education: as criterion for reemployment, 121-22, 161; as method to reduce displacement, 161; *See also* Reemployment

Ehrenberg, Ronald, 134

Electronic components industry: data

of imports on, 158-59; effect on community of, 2; effect on worker of, 2; in industries studied, 43-48, 58-59, 158; by industry, 1-2; by region, 1, 42, 47; and worker characteristics, 142-44; *See also* Displaced workers
Job Training Partnership Act, Title III (JTPA, Title III), 6, 8; *See also* Displaced workers
Johnson, Steve, 28 n1

Klein, Katherine J., 7, 156

Labor: as production cost, 79-80, 81
Labor market: displaced worker in, 128; and local reemployment, 130, 144
Leary, Thomas J., 70
Least squares test: for effect of prenotification on financial losses, 137-44
Legislation, proposed: and displaced workers, 6-8, 133-34
Legislation for displaced workers: in Great Britain; in Sweden; in states; in West Germany, 7-8
Leighton, Julia, 136
Lipsky, David B., 122
Litvak, Lawrence, 49
Loan guarantees: as factor in employee takeovers, 156
Location shift, industry: as component of growth, 34-42; factors related to decision for, 3-4, 65, 67, 103, 105
Location shift, worker, 130
Logit model: to test plant closing rates, 53-57, 60-61; to test probability of plant closing, 78-107
Logue, John, 157

McGranahan, David, 50
McKenzie, Richard, 51, 134
Mack Truck plant closing, 122
Maddala, G.S., 137
Malecki, Edward J., 69
Mansfield, Edwin, 3, 70

March, James G., 50
Market structure: differences in, of industries studied, 92
Markusen, Ann, 3, 4, 92
Maryland Department of Economic and Community Development, 19-20
Massey, Doreen, 67, 105, 106
Materials: as production costs, 79
Meegan, Richard, 67, 105, 106
Mera, John, 74
Metalworking machinery industry: data for, 8-9; employment distribution in, 4-5, 32-34, 152; employment trends in, 9; estimates for plant closings in, 53-59; job loss in, 43-48, 58-59; restructure in, 9-10; selection for study of, 13-14; SIC code for, 13; and tests for probability of plant closing, 78-107
Migration of industry. *See* Location shift, industry
Model: of costs of displacement, 124-33; to predict probability of plant closing, 78-107; to test plant closing rates, 53-57, 60-61
Mondale, Walter (Senator), 7
Motor vehicle industry: data for, 8-9; employment distribution in, 4-5, 32-34; and employment rates, 152; employment trends in, 9; estimates for plant closings in, 53-59; job loss in, 43-48, 58-59; restructure in, 9-10; selection for study of, 13-14; SIC code for, 13; and tests for probability of plant closing, 78-107
Multiplant firms: employment in, 52; and plant closings, 49-53, 153-54; and profitability, 154-55; *See also* Branch plants; Single-plant firms; Subsidiary plants

National Center for Employee Ownership, 156
National Employment Priorities Act; initiatives for, 7

169

New England Provision Company plant
closing, 51

O'Connell, Mary, 122
Odle, Marjorie, 15, 42, 52, 152
O'Farrell, P.N., 102
Output. *See* Production
Outsourcing: as means to lower
production costs, 71-72, 75-76

Packard plant closing, 122
Palen, John, 122
Peterson, George, 32
Phillips, Kevin, 75
Plant closing legislation: federal
initiative for, 6-7, 8; in Great Britain;
in Sweden; in West Germany, 8; in
various states, 7
Plant closings: for branches and
subsidiaries, 10, 49-51, 66; as
component of growth, 34-42; and
demand for product, 77, 87-90; factors
contributing to, 10, 49-59, 65-107; in
Great Britain, 67, 102, 106-07;
hypothetical factors contributing to,
69-107; and import competition, 66,
74-77; industry age as a factor in, 53,
90, 103; and industry variations, 158,
in Ireland, 102; model to test
probabilities of, 78-107; and multiplant firms, 49-53, 153-54; and
production costs, 65, 152-53; and
regional variation, 3-5, 31, 35-39,
47-49, 53-59, 60-61, 151-52; in
Scotland, 102; and single-plant firms,
49-54, 66, 153-54; test of hypotheses
related to, 53-57, 60-61, 92-107; and
unionization, 65, 153; variations in
rates of, 47-49, 158; and wages, 153
Playskool plant closing, 122
Podgursky, Michael, 120, 121, 123
Prenotification, 7-8; and collective
bargaining contracts, 135; and federal
public policy options, 133, 134,

156-57; and financial losses of
displacement, 137-44, 161-62; as it
affects reemployment, 133-142; and
state and local policy, 157
Probability estimates. *See* Logit model
Product cycle: definition of, 72
Product cycle model, 53, 75-76
Production costs: changes in, as variable
in probability model, 70, 79-82,
83-85; comparison among regions of,
65-66, 70-72, 105; as factor in plant
closings, 65, 70-74, 152-53; and
location shift, 3-4; and outsourcing,
71-72, 75-76; and plant closings, 152;
as variable in probability model, 79-82
Profits: and firm ownership, 155; for
single-plant and multiplant firms,
154-55
Public policy: federal policy to assist
displaced workers, 5-8; implications
for federal, 151; options for federal,
11, 156-59; state and local, 7, 157
Public policy, proposed: to assist
displaced workers, 161-62

Quilligan, James B., 157

Race: as criterion for reemployment,
122, 133; *See also* Reemployment
Reamer, A., 67
Reemployment, 115-45; criteria for,
120-129; and local labor market, 130;
and prenotification, 134-44
Rees, John, 68
Regional differences: absence of, in
plant closing rates, 53-59, 105
Regional growth, 34-42, 47-49, 152;
effect of plant closings on, 35-39,
47-59, 152; and employment gains
and losses, 35; and migration patterns,
35; and plant closings, 35, 152
Regression analysis: and employment
growth, 35
Reich, Robert, 51

170

Wage concessions: and plant closings, 157-58

Wages: as factor in plant closings, 4, 153-54; as factor in plant start-up, 3, 68; as production costs, 80, 83, 84; *See also* Production costs

Walker, Richard, 4, 50

Weissmann, Barbara J., 157

Welch, Finis, 123

Wendling, Wayne R., 153

Westaway, J., 50

Wheaton, William, 82

Whyte, William F., 156, 157

Wickwire plant closing, 122

Williamson, Oliver, 49-50

Wood, K. Haydn, 157

Worker characteristics: as criteria for reemployment, 120-23; and probability of job loss, 142-44

Young, Karen M., 7, 156

Zibman, Daniel, 50